The First World War necessitated not
the industrial resources of all the Gre _____ analyses the
political constraints and ramifications of industrial mobilization in Russia. It
focuses on the War-Industries Committees, an organization founded in 1915
ostensibly to assist in the war effort but also incorporating the aspirations of
the Russian industrial bourgeoisie for greater control over the administration
of the war economy and, ultimately, political power.

Drawing on Soviet and Western archives, contemporary newspapers, the
publications of the War-Industries Committees, and a wide range of
monographic literature, the author pieces together the ways in which
industrial mobilization became a highly charged political issue, one that
contributed towards the downfall of the old regime. After an introductory
chapter on the 'making' of the Russian industrial bourgeoisie, the author
examines initial efforts to mobilize industry, the crisis which brought about
the creation of a network of War-Industries Committees, and the challenge
they posed to the Tsarist government. The next two chapters explore the
Committees' involvement in war-industrial production and the regulation of
the war economy. This is followed by a consideration of the workers' groups
and their attempts to put industrial relations on a 'European' footing. A
final chapter discusses the role of both the Committees and their workers'
groups in the social and political upheavals which preceded the February
Revolution, and their respective fates in 1917.

Dr Lewis H. Siegelbaum is a Senior Lecturer in History at La Trobe
University, where he has been teaching since 1976. He has written extensively
on Russian social and economic history, and is engaged on a study of Soviet
workers in the 1920s and 1930s.

List of Tables

Preface and Acknowledgements

The First World War, the Great War as it was commonly called before another cataclysmic struggle engulfed the world, took over 8 million lives and provoked social and political changes far exceeding almost everyone's expectations or fears. What made the 1914-18 war so destructive, and what at least in part accounts for the pressures for social and political change, was the unprecedented mobilization of industrial resources. That is, the very means by which the European powers increased their capacity to destroy each other also contained a threat to the ways of life and the political systems which those powers were attempting to defend.

In this sense, industrial mobilization represented a quickening of the pace of industrial development with all of its attendant dislocations and disruption. Whether the changes provoked by industrial mobilization could be contained within pre-existing structures or whether they would lead to social disorder and political revolution depended on the nature of the pre-existing order and that of the mobilization. No state could afford to rely solely on its coercive powers or its traditional methods to secure the loyalty and/or compliance of its population. Each resorted to an expansion of its bureaucratic apparatus. But entailed in that expansion was the representation of social interests and groups previously outside, or even hostile to, the state machinery. Hence, in France the Third Republic's *union sacrée* was achieved in part by the inclusion of four socialist ministers; the national coalition in Britain was forged in 1915 by the integration of business and trade union organizations into the state's administration of industry; and the near militarization of industry and labour in Germany was achieved by the extraordinary degree of cooperation between military, business and trade union leaders.

Among the Great Powers, this process of coopation was least developed in Russia during the war. Not only did the Tsarist regime fail to accommodate itself to working class representation, but even the middle classes and, particularly, large sections of th« industrial bourgeoisie were denied access to the levers of state power and became rivals to the state's mobilization of industry.

The present study focuses on this rivalry. It analyses the politics of industrial mobilization in Russia through an investigation of the War-Industries Committees. This was an organization of industrialists set up in mid-1915 to assist the state in its belated efforts to put Russia's economy on a war footing. The immediate inspiration for the creation of the committees was the shortage of munitions experienced by the Russian army. This shortage, which had existed since the outbreak of the war, took on crisis dimensions in the spring and summer of 1915 when the Russian armies retreated from Poland and Galicia. The momentum generated by that crisis soon spilled over into other questions of production and distribution. It raised serious doubts within the industrial bourgeoisie about the ability of traditional state contractors to furnish an adequate supply of war materials, the Tsarist bureaucracy to administer the war economy, and the very system of government to sustain the war effort. It thus became a political question, and, in so far as the mobilization of labour power was concerned, a social question as well. Even after the munitions shortage had been overcome – largely through means other than those advocated or represented by the War-Industries Committees – the crisis of confidence generated by the shortages continued, justifying the committees' existence and complicating the mobilization of industry.

The significance of the War-Industries Committees, then, was not limited by their role in the production and distribution of war materials, though in this respect they contributed more than has hitherto been acknowledged. The world of the committees was a microcosm which reflected the most profound issues of Russian industrial life. The state's guardianship of industry, the policies of its regulatory agencies, the development of new products and processes to overcome dependence on foreign technology, the relationship of technical specialists to industrial development, and the necessity to forge links with the working class – these were just some of the concerns raised by the committees. In addition, representing the most politically dynamic elements within the industrial bourgeoisie, the committees were frequently in the forefront of the liberal opposition movement. As such, their fate was intricately bound up with the changing fortunes of the Duma's Progressive Bloc and its campaign to establish a 'government of public confidence' with the February Revolution, and with the Provisional Government in which several of the committees' leaders served as ministers. Finally, the committees' sponsorship of working class representation in the form of autonomous workers' groups added a new dimension to relations between the liberal and social democratic movements, one which in important ways prefigured the experience of 1917.

There has previously been no full-length study of the War-Industries Committees. Soviet historians dealt fleetingly with them in the 1920s and 1930s in connection with the publication of memoirs by ex-Tsarist military and civil officials, and of the bulletins of, and other materials relating to, the workers' groups.[1] This, the inter-war period, also saw the publication by the Carnegie Endowment Fund of a number of studies by *émigré* academics and journalists of a liberal persuasion. These were, in the main, far more sympathetic to the War-Industries Committees and the other major public organizations, the Union of Towns and the Zemstvo Union.[2] Those shortcomings in their activities that were acknowledged were generally attributed to the Tsarist government's refusal to share or devolve its authority.

On the eve of the Soviet Union's involvement in the Second World War appeared the only comprehensive article ever published in the USSR on the War-Industries Committees.[3] The aim of the

article was to demonstrate that the committees'
contribution to the war effort was negligible and
that their main function was the political mobilization of the Russian bourgeoisie against the
rising tide of worker unrest. This argument, which
became the touchstone for many subsequent studies
bearing on the committees, rested on weak ground –
a speech given at the committees' second congress,
a quotation from the chairman of the central
committee referring to tactics adopted by liberals
in 1913, and other slightly misleading information.[4] No archival material pertaining to the
committees was used.

The Soviet victory in the Second World War
invited comparisons with the less successful
performance of the Tsarist state and war-
industrial production in 1914-17. A dissertation
submitted in 1947 by N.I. Razumovskaia analysed
the Central War-Industries Committee's economic
activities in considerable depth and on the basis
of extensive archival material.[5] However, the
author neglected to discuss several important
questions, such as the relationship of the central
committee in Petrograd to the provincial
committees, the preparations for post-war development, the role of technical specialists, and the
issues raised by the workers' groups.

The administration of the economy in the First
World War and its shortcomings attracted a good
deal of attention in the post-Stalin years. A.L.
Sidorov's numerous articles and K.N. Tarnovskii's
study of the metallurgical industry contain many
insights into the conflicts between the government
and the committees and among industrialists.[6]
Much of their work was based on archival material
still unavailable to Western scholars. The
considerable debt which I owe to these historians'
works is acknowledged, especially in Chapters 2
and 6.

In 1967, the fiftieth anniversary of the
February and October Revolutions, two Soviet
historians, V.Ia. Laverychev and V.S. Diakin, gave
the committees prominence in their studies of the
political activities of the Russian bourgeoisie.[7] Diakin's in particular is a character-
istically judicious and closely argued work. But
neither book attempts to link political strategies
and fortunes with industrialists' roles in the war

economy. Of great interest also, as much for the
questions it poses as for the provocative conclusions which it draws, is George Katkov's *Russia
1917: The February Revolution*, also published in
1967.
While Soviet historians have written extensively
on the internal composition of the Russian bourgeoisie, this question has only recently attracted
attention among Western scholars. In particular,
the degree of conflict between Moscow and
Petersburg industrialists has been the subject of
an interesting debate, with one contributor
viewing the Central War-Industries Committee as
'the chief battleground for the struggle' during
the war years.[8]
More recently still, two works have appeared
which deal with issues raised in the present
study. One, by Raymond Pearson, concentrates on
the manoeuvrings of the parties within the
Progressive Bloc. Its chief value is in detailing
the dilemmas of the Russian 'moderates' and their
inability to transcend the 'crisis of Tsarism'.[9]
The discussion of the industrial bourgeoisie and
the War-Industries Committees unfortunately is
marred by a number of inaccuracies. Pearson's
failure to differentiate among 'industrialist
circles', his frequent and puzzling references to
the state's intention to 'nationalize' war industries, and his placement of the Central War-
Industries Committee in Moscow suggest that for
him industrialists and industrial mobilization
were not central concerns.
Norman Stone's book, *The Eastern Front, 1914-
1917*, is another matter entirely. It is, in many
ways, a model of military and political history.
Interpreting the impact of the war 'not as the
vast run-down of most accounts but as a crisis of
growth – a modernisation crisis in thin disguise',
Stone convincingly disposes of a number of legends
concerning the effects of backwardness and military unpreparedness on the war effort.[10] When he
comes to consider the role of the War-Industries
Committees, however, Stone's revisionism (that is,
his attempt to counter the claims of 'their propa-
gandists') goes too far. He argues, for example,
that 'where they succeeded they were unnecessary,
where they did not they were a nuisance', and
concludes that 'the War-Industries Committees

1 Russian Industry and the Making of a Russian Industrial Bourgeoisie

For much of the nineteenth century, Russian industrial development was shaped by the experience of serfdom and then, after Emancipation, by the persistence of feudal relations between landlords and peasants. The state, by allowing the nobility to mortgage serfs for non-productive as well as productive expenditure, by generously compensating it for the loss of serf ownership and by embarking on the construction of railways designed to expedite the grain trade, limited access to capital by industrial entrepreneurs. At the same time, the reinforcement of communal ties through collective responsibility for redemption payments favoured a cottage industry as opposed to a wage labour force. The existence of numerous state-owned enterprises including the State Bank, legal restrictions against national and religious minorities, and the perpetuation of the guild system defining the rights and obligations of the merchant estate, the *kupechestvo*, reinforced these limitations on private entrepreneurial activity and the development of a well-defined or even self-defined bourgeoisie in Russia.

Yet, what the state was doing in one respect it was undoing in another. The means by which the old legal and social orders were being preserved were at the same time undermining them. Not only was the state through its educational institutions producing an intelligentsia increasingly committed to political action against the state, but in attempting to keep pace with European imperialist expansion, it fostered a new cadre within the bureaucracy less committed to traditional

1

practices.   Concentrated   for   the   most   part   in   the
Ministries   of   Finance,   Ways   and   Communications,
and   State   Control,   these   state   officials   bore
little   resemblance   in   terms   of   modes   of   recruitment,   social   affinities,   and   styles   of   work   to
those   charged   with   the   administration   of   civil
order,   state   domains,   and   church   affairs.[1]   By
the   1890s,   when   industrialization   began   in
earnest,   the   balance   had   been   tipped   in   favour   of
the   new   mode.   Under   the   guidance   of   Sergei   Witte,
Minister   of   Finance   from   1892   to   1903,   the   state
extended   massive   amounts   of   credit,   imposed   the
highest   tariffs   in   Europe,   and   gave   every   encour-
agement   to   foreign   investment   in   private   industry
and   state   bonds.   This   programme   produced   annual
growth   rates   exceeding   8   per   cent   in   the   heavy   and
extractive   industries.   It   resulted   in   Russia   being
among   the   world's   leading   producers   of   coal,   pig
iron,   oil   and   several   other   industrial   goods   by
the   turn   of   the   century.   It   also   fostered   the
growth   of   private   enterprise,   enabling   it,   ironically,   to   acquire   a   degree   of   independence   from
the   state   which   it   hitherto   had   lacked.

However,   any   claim   to   extensive   capitalist
development   has   to   be   carefully   qualified.   As   late
as   1917   Russian   society   was   still   overwhelmingly
agrarian   and   despite   a   marked   shift   in   state
policy   after   the   1905   revolution   towards   encouraging   capitalist   farming,   traditional   economic
patterns   prevailed.   Equally   significant   in   terms
of   economic   development   and   class   formation   was
the   continuing   presence   of   the   absolutist   state.
In   Western   Europe   and   the   United   States   the   role
of   the   state   had   evolved   into   one   of   securing   the
general   conditions   of   capitalist   production   by
providing   systems   of   law,   defence,   education,
transportation   and   communication   without   which   the
expanded   reproduction   of   capital   could   not   take
place.   In   Russia,   however,   the   equation   was
reversed.   The   expansion   of   industrial   capital   was
encouraged   by   the   state   as   a   means   of   securing
conditions   favourable   to   an   increase   in   the   military   and   strategic   capacities   of   the   empire.[2]

In   this   context,   the   peculiarities   and   difficulties   in   the   making   of   a   Russian   industrial
bourgeoisie   can   be   readily   appreciated.   Until   the
1890s,   Russian   businessmen   represented   themselves
sectionally   in   merchant   societies,   local   exchange

committees, or in associations uniting branches of industry such as those of South Russian, Urals and Polish mineowners, Baku oil producers and central Russian ironworks manufacturers. No one organization presumed to speak for industry and commerce in general. Occasional all-Russian congresses of industrialists such as those held in St Petersburg in 1870 and 1872 and Moscow in 1882 only tended to reproduce existing sectional interest in their resolutions.[3]

The problem was not so much legal, but the absence of law and its associated infrastructure. As Dmitri Mendeleev, the world-renowned chemist, explained to assembled businessmen in Moscow in 1882,

we, so to speak, are reaping the fruits of the past, when each *chinovnik* was able to treat the factory owner and entrepreneur as the landlord treated the peasant. Factory matters were, it is true, considered with patience, but not really more than as the whim of the entrepreneur, and the factory owner was only able to consider himself free from various restrictions derived from the absence of a clear law when he was wealthy and made gifts.[4]

In other words, far from providing the framework within which businessmen could operate collectively and with confidence, the state permitted only individuals to buy their way into its good graces, or to enhance their stature by becoming ennobled.

In his speeches and writings which spanned the last two decades of the nineteenth century, Mendeleev frequently called on the state to issue 'clear' laws for Russian industry, to promote polytechnical education to provide industry with native-born engineers, and to encourage the publication of journals dedicated to the application of science to industry. But Mendeleev was something of an exception – a non-Marxist *intelligent* fully committed to industrial capitalism as 'the only reliable means of furthering the development of our prosperity...and reconciling the interest of the masses of people with the interests of the educated classes'.[5] The bulk of the intelligentsia, whether Marxist or neo-

Populist, would have little to do with the promotion of the interests of business and in general shared the landed nobility's contempt for the 'practical' (*delovoi*) man. It did not, therefore, provide the intellectual glue which could unite industrialists or make them adhere to other elements to comprise a bourgeoisie, but rather left them to their own devices, which were meagre. The pursuit of narrow economic interests, a survival mechanism in an unsympathetic environment, itself perpetuated such divisions.

That environment began to change in the 1890s, though no thanks to Russian businessmen or the intelligentsia. The really dynamic elements within Russian industry in the late nineteenth century, as many studies have demonstrated, were either state officials such as Witte or foreign entrepreneurs. The 'Russian Rockefellers' were not Russian at all, but the Nobel family which, although resident in Russia for three generations, retained numerous ties with its native Sweden. The pioneers for profit, so impressively portrayed by John McKay, were from France, Belgium and to a lesser extent Britain and Germany.[6] Backed by the leading European financial institutions but in some cases also striking out on their own, they were largely responsible for the establishment of the South Russian coal, iron ore and metallurgical industries and figured prominently in banking, the metalworks industry and electrical power generation. By 1900 over 70 per cent of the capital in the heavy and extractive industries was non-Russian.[7]

Producing mainly for the state's railways and armed forces as well as their own productive consumption, these industries were organized in a complex network of syndicates, trusts and less formal marketing arrangements. In structural terms this process was similar to what was taking place or had already occurred in Germany, France and the United States. But the relative absence of native capital, the vagaries of Tsarist law, and the inexperience and volatility of the Russian work force presented a unique combination of opportunities and risks. In the case of Prodamet, the ferrous metals syndicate formed in 1902, and Produgol', the coal syndicate founded in 1906, French banks and their representatives in Russia played an early and decisive role.[8] During the long recession which these industries experienced between 1900 and 1910, the syndicates consolidated their positions by reducing commercial costs and outbidding the 'outsiders'. By 1910 the 24 enterprises comprising Produgol' produced 64.4 per cent of all coal (excluding anthracite) in the Donets Basin, and in 1912 Prodamet sold 85 per cent of all the products of Russian metallurgy.[9]

In the meantime, native-born engineers and managers, many of them ex-state officials, were taking their places alongside Russianized foreigners in the upper echelons of the heavy industrial enterprises. This process, part of the merging (*srashchivanie*) of state and private entrepreneurial interests which several Soviet historians have detected, occurred in conjunction with a shift towards passive portfolio investment on the part of foreign capital and the expansion of technical training facilities in Russia.[10] It may also have stemmed from the difficulties which foreign managers and engineers experienced in dealing with Russian workers.[11]

Just as retired bureaucrats figured prominently on the boards of the leading banks and heavy industrial enterprises, so a number of businessmen participated in inter- or sub-ministerial committees. Nevertheless, as demonstrated by the state's experimentation with police unions and industrialists' flirtations with oppositional politics during 1905, the process of *srashchivanie* was only in its early stage.

The revolution of 1905, revealing both the tenuousness of the autocracy's hold over society and the dangers which mass action posed to the economic order, convinced even the least politically inclined businessmen of the necessity to represent their interests before the state and the rest of society in some collective fashion. But, as suggested by the failure of several congresses of businessmen to generate a common programme during the first half of the year, interpretations of what those interests were and how they could best be articulated varied considerably. The October Manifesto, drafted by Witte and reluctantly signed by the Tsar, provided a new opportunity for such representation. Its promulgation was the signal for the establishment of several political parties, some of them short-lived, and each of them jostling for support among industrialists. With the liquidation of the Soviet of Workers' Deputies in Petersburg, the crushing of the Moscow uprising, and the restoration of order in the provinces, the more radical elements within the middle class were put on the defensive. The 'government of national confidence' which the Kadet (Constitutional Democratic) Party and the Moscow-based Moderate Progressive Party had sought, proved to be chimerical. The dismissal of the first two Dumas and the revision of the election laws finally brought the revolutionary crisis to a close and ushered in a working relationship between the government and the uneasy alliance of landed and conservative industrial interests represented by the Union of 17 October, better known as the Octobrists.

By this time, most Russian businessmen were eager to return to business as usual, and it was to this desire that the Association (*Sovet s"ezdov*) of Industry and Trade, founded in 1906, addressed itself. The nature of the Association and the views which it advanced have been analysed by several historians in a number of contexts.[12] Here it will suffice to note that from its inception the promotion of industry, or what was referred to as 'the development of Russia's productive forces', predominated over trade in its deliberations and representation and that within industry syndicalized mining and metallurgy commanded the greatest attention and support. Translated into practical measures, this meant convincing the appropriate ministers to ease credit and reduce taxes, to maintain high tariffs, encourage foreign investment, regularize the system of distributing state orders, and limit state entrepreneurship.

The Association's position was formulated and defended by its 'organization men', those who had accumulated experience reconciling diverse interests among enterprises and their financial backers and between different branches of industry and the state. Perhaps the most outstanding in these respects was N.S. Avdakov, an Armenian-born engineer. For many, both within and outside the organization, Avdakov, who served as chairman from 1906 until his death in 1915, *was* the Association.

He was also a founding member of Prodamet, chairman of Produgol', commercial director of the Société Générale's Rutchenko Coal Company, a member of the upper legislative chamber, the State Council, and for a brief period in 1915, chairman of the Central War-Industries Committee.

Among the diverse interests which the Association claimed to represent, the textile industry, concentrated in the central industrial region around Moscow, occupied a somewhat ambiguous position. To understand the role which the Moscow textile manufacturers played within the industrial bourgeoisie and their ascendance to national political prominence during the war, a brief account of their origins, their business activities, and the issues with which they were involved in the decade before the war is in order.

The mechanized production of textiles in Moscow dates from the early years of the nineteenth century. But it was not until the 1840s, when a number of cotton spinning enterprises were established by members of the Old Believer religious communities, representatives of the lower urban orders (*meshchanstvo*), and serfs who had bought their freedom, that it became Moscow's chief employer of industrial labour.[13] By 1858, textile workers in Moscow numbered 34,400 or over 80 per cent of all industrial workers registered in the city.[14] With the expansion of the printing, food processing, leathergoods and above all metalworks industries in the latter half of the century, the relative weight of textiles diminished. Still, textile workers continued to comprise the largest sector of Moscow's factory population, accounting for 42 per cent of all factory workers in 1913. This was, however, only a quarter of the total number employed in textiles in Moscow *guberniia* (province).[15] The majority lived and worked in such industrial towns as Bogorodsk, Kolomna, Serpukhov and Orekhovo Zuev. There were also major concentrations of cotton, woollens and linen factories further afield, in Vladimir, Kostroma and other provinces of the central industrial region.

The textile manufacturing families, which showed a remarkable degree of continuity throughout the nineteenth and into the twentieth century, typically both owned and managed their enterprises,

relied on locally generated capital, and
distributed their products through outlets in
nearly every town in the empire.[16] Despite the
geographical dispersion of their enterprises, the
manufacturers developed a strong corporate
identity which was reinforced through their domination of the Moscow Exchange Committee. They also
occupied an important position in Moscow's public
life. Thus, in the popular imagination Holy Moscow
eventually gave way to 'calico Moscow', the manifestations of which were not only the large cotton
mills and massive quantities of cotton goods
passing through its freight yards, but also the
sumptuous if sometimes bizarre domiciles of the
manufacturers, their involvement in municipal
administration, their patronage of the arts,
participation in (Old Believer) religious affairs
and philanthropic endeavours.[17]

As the pace of industrial and financial activity
quickened in the 1890s, the Moscow textile manufacturers were increasingly drawn into questions
of national economic policy. They were dependent
on, and clearly benefited from, the state's maintenance of law and order, railway construction,
tariff protection and, later, Stolypin's agrarian
reforms. They were, none the less, relatively
distanced from the bureaucracy which, as has
already been pointed out, had other reasons for
pursuing such policies. Thus, an unidentified
textile manufacturer could write to the journal
Promyshlennyi mir (Industrial World) in 1900:

> In other countries they also have authorities
> for the affairs of trade and industry, but they
> have a different function. There they manage,
> i.e. satisfy the needs of industry and trade,
> and do not command those needs. This is a
> distance of huge measure, but all the same, it
> is necessary for us to bridge it.[18]

With respect to the increasing weight of foreign
capital in Russia,the Muscovites were not over
enthusiastic. In January 1899, the Moscow Exchange
Committee adopted a resolution drawing the government's attention to 'the damage to Russian
interests and the danger represented by the
further expansion of the activities of foreign
enterprises in Russia'.[19] The resolution,

Vyshnii-Volochek, the Riabushinskii family owned
and managed a bank which it had acquired in 1900.
Between 1908 and 1914 the bank established ten
branches in the flax growing region to the north
and west of Moscow. In 1912 it was converted into
the Bank of Moscow, operating as a joint-stock
company. Through the bank, the Riabushinskiis
acquired a controlling share in RALO, Russia's
leading linens manufacturing and export company.
They also purchased a paper mill which supplied
the family's printing press.[31]

Diversifying their interests, the Riabushinskiis
- and other manufacturers - sought to expand their
share of foreign markets, particularly those lying
to the south of the empire. Towards this end, the
Russian Export Company was formed in 1908. Within
four years, Russian fabric exports doubled and by
1912 they surpassed Britain as Persia's largest
supplier of cotton goods.[32] To what extent these
commercial advances alerted Russia's industrial
bougeoisie to the dangers of German penetration
(manifested in the Berlin-Baghdad railway) and
predisposed it towards military conflict, is
difficult to determine. As argued above, the sort
of militant liberalism expounded by Moscow's
manufacturer-Progressists had various roots. Among
these was the sense of national disgrace derived
most obviously from Russia's defeat at the hands
of the Japanese in 1904-5 and the 10-year commercial treaty signed with Germany in 1904 which was
widely regarded as detrimental to Russia's industrial development. The blow to Russia's prestige
as a result of the annexation of Bosnia-
Herzegovina by Austria in 1909 only intensified
this sentiment. The subsequent allocation of large
sums for naval reconstruction and the reorganis-
ation of the army did not alleviate it.

National liberalism by no means had a monopoly
on concern about Russia's commercial and strategic
position *vis-à-vis* the other Great Powers. The
Neo-Slav movement which arose at this time among
sections of the Polish and Czech intelligentsia
enjoyed the active support of Octobrists and the
right-wing Nationalist party. In the Third Duma,
Guchkov launched stinging attacks against the
incompetence and nepotism which he held responsible for the military disasters during the
Japanese War.[33] But it was the national liberals

with whom the idea of a 'Great Russia' became most closely associated, revealing a strong militaristic streak in their ideology.

This was nowhere so apparent as in a two-volume collection of essays entitled *Velikaia Rossiia*, which appeared in 1910-11 in a lavishly produced edition by the Riabushinskii press. Dedicated to what its editor, V.P. Riabushinskii, called 'the ideological preparation of the country for battle' and a 'healthy militarism', the collection contained contributions from such liberal intellectuals as P.B. Struve and Professor S.A. Kotliarevskii - both of whom had participated in the 'economic discussions' referred to above - as well as experts in military and naval science.[34] The identification between Russia's commercial-industrial development and its greatness was stressed by several authors. Struve, writing on 'The Economic Problems of Great Russia', saw them being overcome by 'the spirit of national Europeanism - built by free men on liberated foundations with healthy competition'.[35] Another contributor, L.M. Iasnopol'skii, argued that

> Only the furthest extension of Russia's internal market and above all commercial-industrial credit and state capital, the intensification of economic activity and the differentiation of branches of the economy - all that in general characterizes the transition of a state from agrarian to industrial - this is what is necessary for providing the domestic basis to carry on a great war.[36]

That such a war was inevitable and that Germany would be the likely enemy was predicted by *Utro Rossii* as early as October 1911. The paper urged the government to strengthen its ties with France and Britain so that it could concentrate in the meantime on 'applying the creative powers of its national policy towards the East'.[37]

Thus, after 1905, the formation of the industrial bourgeoisie in Russia had proceeded along two distinct lines. One involved the gradual Russification of management within the heavy and extractive industries, the horizontal and, to a lesser degree, vertical integration of their enterprises into a variety of syndicates and

trusts, and the explicitly non-political representation of the interests of industry and
commerce through the Association of Industry and
Trade. The other followed a course which repeated,
in some respects by conscious imitation, the
classical pattern of bourgeois class formation
in western Europe. Based primarily on mass
consumption and domestically generated capital, it
represented the transformation of the juridically
defined *kupechestvo* into a national class with
aspirations for political leadership.

But in the years immediately preceding the
outbreak of war, these trajectories were altered
by a number of developments which have been
variously interpreted as growing pains associated
with the process of modernization, and as the
onset of a new and potentially more explosive
revolutionary crisis.[38] Thanks to a surge of
state demand in connection with its strategic
railway and armaments programmes and the restoration of foreign investor confidence, heavy
industry experienced a boom after 1909 which
placed fuel and raw materials suppliers in an
awkward position. Efficient at restricting
production and competition during the long
recession which preceded the boom, the monopolistic organizations dominating these industries
adopted a variety of strategies to meet the new
demand.[39] In the case of the oil industry,
effectively controlled by the Nobel-Mazut and
Rothschild (after 1912, Shell) groups, this meant
stabilizing output at or below levels achieved at
the turn of the century, doubling prices between
1910 and 1913, and justifying such measures in
terms of the exhaustion of the Baku wells.[40] One
of the effects of this policy was to intensify the
demand for coal by the railways and industrial
enterprises. As Produgol's control over production
was indirect and in any case covered only about
half of total output, there was greater opportunity for individual mining companies to expand
production. Between 1910 and 1913 coal production
rose by 43.3 per cent and in the Donets Basin by
51 per cent.[41] These increases were not,
however, sufficient to satisfy demand. The result
was the phenomenon picturesquely referred to as a
fuel 'famine' which in turn was held responsible
by Prodamet for a famine in iron.[42]

The state's response to this situation was
temporarily to waive tariffs on imported coal, oil
and iron ore, to initiate steps towards the
exploitation of oil-bearing land by the naval and
transport ministries, and to take up the case of a
constituent enterprise of Produgol' which was
attempting to breach the syndicate's quota system
for production.[43] The Association of Industry
and Trade rushed to the defence of the syndicates
and private enterprise in general, but discovered
that it was alone. Its journal, under the headline
'The Threat to Private Industry', complained that

> Nowhere in its attempts to expand state enter-
> prise does the government encounter opposition.
> The legislative organs? But they incite it in
> this direction. The press? But in general and
> with few exceptions it is prepared to applaud
> when any measure hateful to industry is adopted.
> The broad circles of (educated) society? Unfor-
> tunately, they are still too indifferent to
> economic questions. All this raises fear for the
> future of private industry.[44]

The eighth congress of the Association, held in
May 1914, was pervaded with such pessimism
relieved only by uncharacteristically strong
condemnations of the government. Avdakov, in
opening the proceedings, set the tone by announcing that 'recently industry had become tense,
regarding the future without faith, as if
expecting each day to bring some new measure
hindering its development'. Both he and V.V.
Zhukovskii, a Polish engineer who was one of few
from the Association's committee to sit in the
Duma, raised the spectre of a flight of foreign
capital.[45] Others spoke of the instability of
industrial life, a condition which the congress'
resolution attributed to 'the government's persecution of private industry'.[46]

But it remained for the delegates from Moscow to
level the sharpest criticisms against the government. Jules Goujon, a Frenchman by birth but for
many years the proprietor and later managing
director of the Moscow Metalworks Company,
declared that 'neither the Ministry of Trade and
Industry nor the Ministry of Finance has any
understanding of the needs of trade and industry'.

Thus, 'in questions of finance, state contracts, and labour we are completely under the yoke of the Ministry of Internal Affairs'. 'If this continues', he warned, 'then I feel, gentlemen, that we will not be masters of our enterprises, that instability will reign everywhere, and if this continues for long, then all industry will go up the chimney.'[47] Not to be outdone, P.P. Riabushinskii told the congress:

> Our government is not talented. If this goes on, then even the broad masses will lose respect for, and faith in, authority. This will be sad, this is intolerable, this can lead to unfortunate consequences. ...A blind state. An orphaned people. One can only hope that our great country will outlive its petty government.[48]

Riabushinskii's warning about the masses' loss of respect for authority and the Association's allegation of persecution, suggest that both fractions of the industrial bourgeoisie were losing confidence in their respective strategies. If the Association had little to show for its efforts to ensure pro-industrialist policies within the bureaucracy, then the Moscow manufacturer-Progressists had made scant progress in their attempts to achieve 'political hegemony'. The persistence of mutual antagonisms between the Kadet and Octobrist parties sabotaged the Progressists' mandate to bring about the unity of all liberals.[49] As the party in the middle of the verbal crossfire, the Progressists more often than not caught flak from both sides.

Overshadowing all else, however, was the upsurge of worker unrest and political protest after the Lena goldfields massacre of April 1912. Strike figures compiled by factory inspectors, though incomplete and somewhat arbitrary, reflect this growing instability (as shown in Table 1.1).[50]

The increase in working class militance provided unwelcome proof to the manufacturer-Progressists that, to cite one of Konovalov's Duma speeches bearing on the issue, 'the police perspective' with which the 'state organism' viewed industrial relations made the strengthening of Russia's productive forces impossible.[51]

Under these circumstances, Konovalov took the

campaign for mobilizing oppositional forces
outside the Duma and beyond bourgeois circles.
Together with P.P. Riabushinskii, S.M. Tretiakov
and N.D. Morozov he set up an Information
Committee in March 1914. Representatives from the
Kadets, Mensheviks, Bolsheviks and non-party
socialists were invited and attended some of its
informal sessions. The purpose of the committee,
as its name implies, was to keep each of the anti-
government groups informed about the activities of
the others, but the prospect of joint action
towards a 'super-organic' (that is, revolutionary)
solution to the crisis in Russian society was also
discussed.[52] Through I.I. Skvortsov, a Moscow
Bolshevik, Lenin attempted to obtain funds from
the committee to meet the costs of the Bolsheviks'
forthcoming party congress, but it is doubtful
whether he succeeded. In any case, Konovalov's
enterprise illustrates the extent to which the
manufacturer-Progressists were willing to go to
increase their political leverage at a time when
all else appeared to have failed. It also
represents an interesting precursor of the War-
Industries Committees' campaign in 1915 to include
worker representation.

TABLE 1.1 *Strikes in Russia, 1910-14*

	Number of Strikes			Working days lost
	Total	*Economic*	*Political*	*(in thousands)*
1910	222	214	8	256
1911	466	442	24	791
1912	2032	732	1300	2376
1913	2404	1370	1034	3863
1914	3534	969	2565	5755

Hence, already before the war the structures
within which various groups would define their
reactions to the war and its tribulations had been
erected or were taking shape. Within a few decades
Russia had simultaneously passed through the age
of coal and iron and was well advanced in the
second industrial revolution, that of electricity,
oil and steel. Concomitant with this development
was the formation of an industrial proletariat
retaining at least in part strong ties to the

land, and a bifurcated industrial bourgeoisie. Far from being merely retrospective analytical categories, these were groups which were becoming increasingly conscious of their own identities, even while the state only grudgingly acknowledged their existence and resisted the articulation of their interests.

The bifurcation of the industrial bourgeoisie was based on different sources of capital, methods of management and market orientations and conformed loosely to the division between heavy and light industries. While it was the former which encompassed the war industries, textile manufacturers based in Moscow were by no means unconcerned with Russia's military preparedness or the fact that it did not measure up to the standards they set for 'Great Russia'. This concern transcended their desire to expand exports. It arose from their ambition to become the nucleus of a national class, a national bourgeoisie, capable of exercising hegemony among other elements of the bourgeoisie and over the rest of society. The war, by severely straining the resources of the state and the methods by which the bureaucracy organized them, gave the Moscow manufacturers the opportunity to represent themselves as the leaders of a popular (*obshchestvennoe*) movement to save Russia. And the issue which precipitated this movement was the munitions shortage and the mobilization of industry.

2 Russian Industrialists and the Initial Mobilization, 1914–15

A few days after Nazi Germany had launched its invasion of the Soviet Union, the BBC's Overseas Transmission broadcast a talk by Sir Bernard Pares entitled 'Democracy Marches'. Pares spoke of an earlier invasion of Russia, one which he had witnessed.

> I was with the Russian armies throughout that time, and I know what the effect was for us at the front. On 2 May 1915, I was at the spot where Mackensen first broke through in the great drive which swept us out of Galicia and far back to the Pinsk Marshes where the Russian Army is fighting today....Men can die where they stand but they cannot beat metal if they have no metal of their own. The Division with which I was that day – normally 16,000 – was reduced to 500; the Regiment from 4,000 to 41.[1]

Pares was referring to the retreat of the Russian armies from Galicia and Poland which had commenced in April 1915 and was only halted in late September. This retreat was a watershed in many respects but none so obviously as in the relationship of Russia's industrial bourgeoisie to the war effort and the Tsarist government's direction of it. The retreat marked a transition from what has been called patriotic enthusiasm to widespread patriotic anxiety based on fear of defeat and its social and economic consequences. While this change did not reach revolutionary proportions, the events of this period had a considerable impact on the situation which confronted the government in February 1917 and the ease with which it was overthrown. One of these events was the formation of the War-Industries Committees. This chapter will assess the relationship between organized industry and the government in the first nine months of the war, up to the Galician retreat and the political crisis which it precipitated. It will focus on the sources and manifestations of industrialists' patriotic enthusiasm, on the munitions shortage in 1914-15 as the principal element in the undermining of this enthusiasm, and on the organizational efforts to overcome the shortage.

2.1   THE PATRIOTIC ENTHUSIASM OF RUSSIAN INDUSTRIALISTS

Among the manifestations of popular support which greeted the Tsar's declaration of war on 19 July 1914, those emanating from business circles were unanimous in their effusiveness and optimism. Sentiments which only weeks earlier would have been dismissed as naive gushed out of the daily press, echoed in the halls of the Tauride Palace where the Duma held its 'historic' one day session, and were repeated in the trade journals. 'There are in Russia neither Rights, Lefts, government nor society, but only a united Russian nation', declared *Utro Rossii*.[2] 'Thanks to the industrial boom and good harvests in the last few years', claimed Petrograd's *Birzhevye Vedomosti* (Exchange Gazette), 'we are completely prepared for, and can withstand without serious upheavals, a protracted war.'[3]

The general mobilization of July-August 1914 did, of course, entail certain economic dislocations. There was a suspension of stock trading activity, a moratorium on debts, the curtailment of most foreign commerce, the reduction of much internal trade, and a loss of manpower. Industrial production fell immediately. The manufacture of textiles dropped 50-60 per cent in the first weeks after mobilization.[4] The prohibition of the sale of spirits not only deprived the state of one of its main sources of revenue, but brought most distilleries to a halt. Coal output in the Donets Basin decreased from 120 million puds in June to 90 million in July and to 80 million in August. This was still too much for the railways which could only transport 44.3 million puds of coal in July compared to 75.5 million the previous month.[5] The situation was made all the more critical by the German blockade of the Baltic and the seizure of the Dombrowa Basin. The Donets would now have to supply Petrograd, previously dependent on British coal, as well as its traditional clients. Most of Petrograd's metal-works factories were operating on only a few weeks' supply of fuel in July and two had no reserves at all.[6]

On the other hand, the actual process of induct-ing, equipping and transporting nearly 5 million men was executed far more smoothly and with less political protest than the government and those who rallied behind it had feared. Whether the sparsity of draft evasions and anti-war activity meant that the vast majority of workers and peasants shared the patriotic enthusiasm of the middle classes, or whether it simply reflected their resignation in the face of widespread arrests of those who were attempting to organize such protests, is difficult to determine.[7] In any case, the war provided an opportunity to discipline those who remained in the factories, and both the government and industrialists seized it. Deferred workers risked losing their status if they changed jobs. Many trade unions were deregis-tered without any objections from employers, some of whom had argued for their necessity before the war. In the Donets coal mines, compulsory over-time, three shifts and the recruitment of women, children and, later on, Chinese workers, compen-sated for the loss of approximately one-third of the labour force.[8] Strike activity, which had reached an unprecedented level on the eve of the war in connection with the struggle among Baku oil workers for better housing and police repression of sympathy meetings in Petersburg, was almost non-existent in the latter half of 1914.[9]

But irrespective of actual conditions, Russian industrialists remained confident about the ability of the economy to sustain the country during the war. The Association of Industry and Trade exhibited this faith, bordering on obscuran-tism. In an article which appeared in its fort-nightly journal, the Association proclaimed:

it is impossible to doubt for one minute that Russia will cope with this war and all its difficult economic consequences more easily than any other nation.

The reasons for this were that 'we have in abundance all goods of primary necessity, especially food. In fact, we are not only provided with them but possess great surpluses.' Those who were concerned about the curtailment of foreign trade could rest assured that 'our country is so large and varied that internal turnover will more than compensate the loss of the external'.[10] At least one foreign economist shared the Association's views and was quoted, to wit:

Of all European countries, Russia in the economic sphere has least to fear from the war. This is due to several factors - large territory and severity of climate preventing military invasion, abundant population, etc.[11]

Such confidence was self-deceptive. It harkened back to the doctrine of the 'less developed the more defensible', which, ironically, had been advanced years earlier as an argument against the development of war industries.[12] It reflected the belief, shared by the military, that mobilization only involved the calling up of troops, that wars were fought on the basis of supplies available at the time of mobilization, and that the responsibility of industrialists was to ensure that business was conducted as usual. It therefore failed to take into account the role of technology, of war-industrial potential.

Yet, sustained by such confidence, industrialists ardently looked forward to the possibility of victory and its fruits. As the Association's journal affirmed, 'There is no sacrifice which Russian industry would not be willing to make', because 'a victorious Russia will be able to dictate its own economic programme.'[13] High on the list of priorities in that programme, indeed 'a fundamental question of the state's existence', was Russia's control of the Bosphorus and the Dardanelles.[14]

The Association of Baku Oil Producers greeted the war with near despair, warning of 'sad consequences' in the form of 'economic upheavals and perturbations' greater than those experienced during the recent strikes in Baku.[15] But a month later, its journal noted that Germany's 'hegemony on the international market' had suffered a 'death blow'. 'Russia', it continued,

> can and must use the war to make its industry flourish. Legislative institutions and commercial-industrial organizations and firms must direct every effort so that raw material sent abroad before the war is processed in Russia. They must find all measures to fabricate at home finished products (hitherto) obtained from abroad.[16]

The journal of the Society of Factory and Mill Owners of the Moscow Region asserted that 'the victory of Russia in this struggle...will give it the possibility...to enter the markets which until now have been held by Germany and Austro-Hungary'. The Society sent a telegram to I.L. Goremykin, the chairman of the Council of Ministers, promising in the name of its 658 member enterprises full cooperation and whatever sacrifices were necessary to obtain victory.[17]

Assistance of another kind was offered by the Association of Urals Mineowners. A memorandum sent to the Minister of Trade and Industry in September 1914 expressed the organization's readiness to distribute orders for munitions. It advertised the Urals as an area with a wealth of untapped raw materials, an experienced labour force, and a location safe from enemy occupation.[18] This was the first offer extended by organized industry to expand the production of munitions and is significant as such. The failure of the government to respond to this and subsequent proposals was eventually to undermine industrialists' patriotic enthusiasm. It also contributed to the already serious munitions shortage.

2.2 THE MUNITIONS SHORTAGE OF 1914-15

The success of mobilization in July-August 1914 obscured Russia's basic unpreparedness in the supply of munitions. Based on estimates drawn up in 1910, the General Staff, a department within the War Ministry, projected that in the event of a full-scale war with a European power Russia would have to field an army of 4.5 million troops at least during the first stage of combat. In terms of rifles, it put production requirements at 4.5 million plus 700,000 per year thereafter, or 6.6 million for 3 years of war. In reality, the army required 5 million for mobilization, 5.5 million to arm those called up later, and 7.2 million to make up losses, or a total of 17.7 million which was 11.1 million more than anticipated.[19]

The projection for machine guns was 4990. Even in July 1914 the army was 883 short and for 1 year of war the demand was 31,170.[20]

The 1910 figures for the annual production of cartridges was 550 million rounds for rifles and machine guns. This left the army 300 million rounds short at the outbreak of hostilities.[21] The average monthly wartime requirement turned out to be 250 million (3 milliard per year) or six times the anticipated amount.[22]

But 'the greatest and saddest influence in the progress of military events', according to General Gurko, the last Chief of the Imperial General Staff (November 1916-February 1917), was the shortage of artillery and especially shells.[23] In the view of the Soviet economic historian, A.L. Sidorov, 'the reorganization of the army carried out in 1910 least of all concerned artillery'.[24] On the eve of the war, the Russian army possessed 7088 artillery pieces consisting of 6336 3-inch field and mountain guns, 512 light howitzers, and 240 heavy field 4- to 6-inch howitzers. The German army had 12,476, nearly twice as many pieces, including 1394 heavy artillery guns of which 996 were the siege type which Russia entirely lacked.[25] Such a disparity was reflected in the relative division strengths of the Russian and German armies. The German division contained 14 artillery batteries of 6 guns each or a total of 84 guns. The Russian division was backed by 7 batteries, each containing 8 guns, or a total of 54 guns. Compared to the German army's 381 heavy batteries, the Russian army contained 60.[26]

To supply this small quantity of artillery with ammunition, the General Staff in 1910 projected an output of 500,000 light shells per month. But this estimate was based on the experience of the Russo-Japanese War. Whereas the Russians expended 918,000 light shells in the course of that war, they used 2.3 million shells in the first 5 months of the Great War. The demand for shells was 500,000 in July but 1.5 million in September.[27]

In short, the Russian army's mobilization plan did not correspond to the reality of war. But the plan itself bears only part of the responsibility for the munitions shortage of 1914-15. Countless monographs and memoirs have demonstrated that all the major belligerents in the war were unprepared for its severity and prolonged nature. Yet the problem was compounded for Russia by a number of factors, among which the meagreness of its armaments industry, contending interests within its bureaucratic structure, and the strategy adopted by those responsible for military procurements in the first months of the war will be singled out for discussion.

As suggested in the previous chapter, Russia's integrity as a great power was a primary consideration in the Tsarist state's promotion of industrialization. This motive was reflected in the type of industry which the state encouraged - heavy over light - the emphasis which it placed on railway construction, and particularly its programme for building strategic lines capable of transporting troops to its western borders. In this light, the retarded and relatively slight development of the domestic armaments industry appears anomalous.

The backwardness of Russia in terms of military technology and productive capacity has been cited as a crucial factor in the Tsar's peace and disarmament initiatives at the turn of the century. Such a condition may be attributed not only to the continued advancements of Britain, France and Germany, but the perpetuation of practices and attitudes within the military bureaucracy which militated against improvement in this area. Among these were the appointments of Grand Dukes to high-ranking positions, the emphasis placed in military training on parade ground efficiency, and the failure to appreciate the significance of improvements in firepower.[28]

Since the Miliutin reforms of the 1870s, the War Ministry had pursued a policy of relying on

domestically produced armaments, obtaining the
blueprints and prototypes for new equipment from
abroad. From the time of Peter the Great until
well into the nineteenth century state-owned
enterprises were the chief suppliers. The strategic principle behind such a policy was obvious
but the small capacity and antiquation of many
plants tended to undermine it. In the case of
small arms production, the army was supplied by
three factories at Tula, Sestroretsk and Izhev;
cartridges and shells were produced by the Tula,
Lugansk and Petersburg factories; the Petersburg,
Obukhov and Perm factories produced artillery
pieces; and five additional factories turned out
explosives and gunpowder.[29] Furthermore, as
large private engineering firms began to win
military contracts, the policy lost much of its
rationale. Not only did these enterprises depend
heavily on foreign finance capital, but foreign
technology in the form of machinery, designs and
engineers remained crucial ingredients in their
military production right up to 1914. The case of
Putilov's successful bid in 1899 for the
production of the 3-inch rapid firing cannon is
illuminating. Having been chosen over several
foreign and state-owned enterprises in a dubious
competition, Putilov sent a delegation of engineers to France to consult with Schneider-Creusot -
one of the original competitors - to improve on
their prototype.[30]

During the Japanese War, the War, Naval, and
Ways and Communications ministries were forced
temporarily to abandon their policy and turn
directly to foreign suppliers. The value of orders
placed abroad by the various departments of the
War Ministry jumped from 2.3 million rubles in
1903 to 16.9 million in 1904 and 73.1 million in
1905. Those for the Navy increased from 10 million
to 68 million in the same period.[31] Figures
cited by General Kuropatkin suggest that at least
for 1904 there was a great disparity between what
was ordered from both foreign and domestic suppliers and what was received.[32]

After 1905, in connection with the state's
rearmament programmes, domestic productive capacity was expanded. While state factories continued to monopolize the production of small arms,
private enterprise garnered an increasing pro-

portion of artillery and shell orders. The share
of war materials in Putilov's annual output grew
in terms of value from 26.8 per cent in 1905 to
42.2 per cent in 1910 and 45.8 per cent in 1912.
This was largely at the expense of locomotive and
railway car production which dropped from 54.1 per
cent to 29.6 per cent over the same period.[33]
Putilov was far from being the only privately
owned armaments producer. By 1914 it was linked
with a number of others under the umbrella of the
Russian-Asian Bank, while the Petersburg
International Commercial Bank provided capital and
directors for another group of enterprises.[34]
The two cartels went to considerable lengths to
influence the appropriate officials within the War
and Naval ministries and, as demonstrated by the
Vickers proposal in 1911 to construct a large
artillery works, competition could be fierce.[35]
Still, as was the case during the Japanese War, so
again in 1914 the army was compelled to seek
supplies from abroad.

The task of rebuilding Russia's armed forces
after their stunning defeat at the hands of the
Japanese was complicated by the fact that the
prosecution of that war and the suppression of the
revolution had severely strained the state's
financial resources. Temporarily bailed out by the
massive French loan of 1906, the regime was still
required to demonstrate its fiscal prudence to
international creditors, and in the conventional
wisdom of the time this meant balancing its
budget. Hence, for much of the period up to 1914,
the military authorities were at loggerheads with
the Ministry of Finance over their conflicting
responsibilities.

Defence expenditure did rise from 473 million
rubles in 1909 to 581 million in 1913 for the army
and from 92 million to 245 million for the navy.
This did not include the 700 million rubles in
capital grants which comprised the 'little
programme' of 1908-9 and the 'reorganization' of
1910, or the even larger amounts consigned for the
'great programme' scheduled for application
between 1914 and 1917.[36] But it was one thing
for funds to be obtained and quite another for
them to be spent effectively. The competing
demands of the War and Naval ministries produced
much politicking within the Council of Ministers,

at the Court, and in the chambers of the Duma.
They did not produce a rational plan for a
balanced military force.[37]
   The minister with responsibility for determining
the army's requirements and passing them along in
the form of state orders was V.A. Sukhomlinov,
Minister of War since March 1909.[38] But
Sukhomlinov, the author of diverting tales of
military prowess dating back to the Russo-Turkish
War (in which he had participated), had many
battles to fight on several fronts - administrative, strategic, political and personal. Whether
this was because of his recognition of the necessity for change, his promotion of younger officers
of relatively low social standing and rank or his
corruptibility remains uncertain.[39] In any case,
as he later admitted,

   In relation to questions of supplying the army
   in general and the realization of credits, I
   assigned responsibility to my assistant as a man
   standing...in this area more competent and
   enlightened than I. Therefore, I placed General
   Polivanov in charge of the Main Artillery Unit
   until 1912, and then General Vernander until
   1915.[40]

The Main Artillery Unit [hereafter GAU, the
abbreviation of the Glavnoe Artilleriiskoe
Upravlenie] was more elusive than Sukhomlinov
implies. Officially under General D.D. Kuz'min-
Karavaev, it was actually ruled by the Grand Duke
Sergei Mikhailovich, who served as Chief Inspector
of the Artillery. Polivanov respected the Grand
Duke as a knowledgeable artillerist, but suspected
his motives in relation to his recommendations for
the distribution of orders.[41] Sukhomlinov not
unjustly considered him a rival. The result was a
stalemate within the Ministry of War which
hampered its ability to obtain credits from the
Ministry of Finance, or rationally to organize the
distribution of those it did receive.[42]
   Despite these problems and others of which he
was only dimly aware, Sukhomlinov authorized the
publication of an article in February 1914
entitled 'Russia Wants Peace but is Prepared for
War'. The article which appeared in several newspapers, asserted, among other things, that

the lessons of the past have not been forgotten.
In a future war, the Russian artillery will
never have to complain of a lack of shells. The
artillery is supplied with a large equipment and
is assured of a properly organized delivery of
shells.[43]

It was an assertion which Sukhomlinov would very
likely come to regret.
As early as the third week of fighting, the
Chief of Staff, N.N. Ianushkevich, wired Sukhomlinov about the huge expenditure of shells and
the need for more supplies. He referred to this
along with the evacuation of wounded soldiers as
'two terrible nightmares'.[44] Sukhomlinov's
response was what one might expect of a minister
in the rear: 'Couldn't measures be taken to
economize on expenditures?' he inquired. Two days
later he blamed the GAU: 'What can I do with this
terrible department with which I have been
fighting since the very beginning of my appointment as War Minister?'[45]
Under pressure from the Supreme Commander, the
Grand Duke Nikolai Nikolaevich, Sukhomlinov called
together on 9 September representatives of enterprises engaged in munitions production. Subsequent
to this and another meeting under Vernander, the
ministry distributed orders for 6.65 million
shells worth 75.8 million rubles to sixteen
enterprises. Despite the fact that the monthly
requirement for shells was three times what the
enterprises engaged to produce them were capable
of turning out (1.5 million compared to 500,000),
the ministry made no attempt to widen the small
circle of domestic contractors. It advanced 9.8
million rubles to them to raise their production
to 1 million shells per month by the autumn of
1915 and assigned orders for 9 million shells to
foreign firms.[46]
These meetings should be contrasted with one
arranged by the French Minister of War, Millerand,
on 20 September 1914. The French ministry, no
better prepared for artillery and shell production
than its Russian counterpart, was sufficiently
impressed by the battle on the Marne to enlist all
directors of industrial enterprises which, according to a preliminary investigation, could possibly

fulfil war orders. One hundred private firms were
divided into twelve regional groups each of which
had its own apparatus for the procurement of
steel, military contracts and personnel. Production of the 75 millimetre shell was expected to
rise from 13,500 per day to 40,000 and eventually 100,000. The middle figure was actually
reached after 4 months and the latter by January
1916.[47]

The contrast was not simply due to the presence
or absence of foresight, fear of acknowledging
shortages, cost-consciousness or concern for high
quality munitions on the part of the respective
war ministers.[48] For such considerations and the
policies which stemmed from them were themselves
functions of pre-existing structures and practices. Whereas France relied primarily on its
home-grown technology and a relatively skilled
labour force, Russian industry could claim
neither. As late as 1913, nearly 70 per cent of
all industrial machinery purchased in Russia was
of foreign origin, much of it coming from
Germany.[49] The production of lathes, presses,
dynamos and other capital equipment during the war
was thus no less important than, and in fact was a
precondition for, the adequate fulfilment of the
military's requirements. The problem was that the
enterprises possessing the capacity to turn out
such items on a large scale were the same ones
which were on contract to produce munitions.

Sukhomlinov later justified the government's
strategy on the grounds that 'there was no time to
create new factories and to develop domestic
industry. It was necessary, therefore, to utilize
what was present.'[50] But neither Sukhomlinov nor
his subordinates in the War Ministry really knew
what was 'present'. Nor did they make any attempt
to find out. Instead, they repeated the example
set by their predecessors during the Japanese War
of relying on foreign firms to supplement domestic
supplies.

Thus, in addition to the consignment of shells
already mentioned, large orders were placed in
late 1914 and early 1915 with British, American
and Japanese firms for gunpowder, rifles, armoured
vehicles and machinery. The fate of these and
subsequent orders from abroad was to become a
matter of dispute among the allies and provoked

much counter-factual speculation and recrimination
after the war.[51] Owing to their unfamiliarity
with Russian specifications and the fact that
allied firms had all they could do to fulfil
orders for their own armies, even with the best
intentions it would have been difficult to provide
Russia with what it needed. The declining exchange
value of the ruble, the chaotic administration of
the allied procurement agencies (Anglo-Russian
Committee, Commission internationale de ravitail-
lement) and the resulting currency speculation
contributed to the low level of fulfilment.
Instead of the 1 million shells from Britain
scheduled for delivery by September 1915, Russia
received 5000. Of the 2 million tons of freight
expected from Britain in 1916 approximately
700,000 tons arrived. By June 1916 American and
Canadian firms had only provided 875,000 of an
expected consignment of 9.1 million 3-inch shells
and proportionately smaller quantities of cartridges and rifles.[52] What made matters worse
was that thousands of tons of supplies remained
for months on the docks of Vladivostok and
Arkhangel'sk, much of it exposed to the elements,
because of inadequate transport and storage
facilities.[53]

But long before the orders distributed in 1914
were scheduled for delivery and despite the War
Ministry's efforts to keep the supply shortages
secret, Russian industrialists were queuing up for
orders. As General Smyslovskii, an official of the
GAU, later testified before the Supreme Commission
to Investigate the Causes of the Russian Army's
Unpreparedness,

> the two meetings and the absurdly high prices
> for orders served as a signal for a whole stream
> of entrepreneurs flooding the GAU with innumer-
> able proposals for the delivery of every kind of
> article especially shells.[54]

In the words of General A.A. Manikovskii, the
chairman of the GAU after the departure of
Kuz'min-Karavaev, 'the 76 millimetre [3 inch]
shrapnel was the first tasty morsel for which all
the jackals bared their teeth'.[55] What
Smyslovskii and Manikovskii neglected to add,
however, was that most of the jackals went hungry

for the time being.

No further orders came from the Ministry of War after October. Those which had been distributed were scheduled for fulfilment by October-November 1915. The moment was opportune for the Association of Industry and Trade to enter the munitions arena. Before the war, the Association had limited its pronouncements on arms production to the observation that 'the first rule of military science must be the preparation of all necessary arms in one's own country', and some reflections, on the centenary of the victory at Borodino, on what constitutes a powerful nation in the twentieth century.[56] But in January 1915, having organized a meeting at the request of representatives of large metalworks enterprises, it drew up a memorandum to the Council of Ministers which contained proposals for the improvement of munitions supply.[57]

Procedures for the distribution of orders were to be reformed. Large factories would receive all orders from the GAU and in turn, when necessary, subcontract to smaller enterprises. To standardize shell production, to control information appearing in the press about orders, and otherwise to ensure the fulfilment of contracts, a special council (*Osoboe soveshchanie*) was to be formed, consisting of government officials and representatives of industrial organizations. Thirdly, all workers previously employed in shell producing factories who had been called to the army were to be returned, and along with engineers and other employees placed under a condition tantamount to service in the armed forces. Finally, steps were to be taken to include in war-industrial production factories 'capable of working for defence', such as those in the Urals.

These proposals have been correctly cited as foreshadowing the Special Council created in mid-May.[58] What has been ignored is the extent to which by advancing them, the Association, an organization supposedly representing all branches and geographical concentrations of industry, was fulfilling the programme of enterprises already engaged in munitions production, among which Petrograd-based firms were predominant. So long as they alone retained direct contact with the GAU, the extension of war-industrial production to other enterprises or areas would be on their terms, regardless of the composition of the proposed special council. This was, then, a plan to consolidate or 'syndicalize' relations among military contractors at a time when it appeared that traditional practices would no longer be sufficient to supply the army or prevent competitors from entering the field.

The ministers, however, reacted coldly to the memorandum. Only I.K. Grigorovich, the Minister of the Navy, considered it useful.[59] S.I. Timashev, the Minister of Trade and Industry, filed a counter-memorandum and reported Sukhomlinov's remarks that the matter was 'already receiving a principled resolution'.[60] The Association's attempt at a pact between industry and government failed.

The principled resolution mentioned by Sukhomlinov took the form of a Special Commission of the Artillery Unit under Grand Duke Sergei Mikhailovich. Created in January 1915 and liquidated less than six months later, the Special Commission did much during its brief existence to expedite the production and delivery of munitions. It encouraged state investment in gunpowder and explosives factories,[61] and exerted pressure on foreign suppliers of shells.[62] Domestic shell production, the Commission's special concern, rose from 229,400 in January to 989,000 in June. But as fast as Russia's factories could turn out shells, the army expended them. Deliveries by June almost equalled the anticipated amount but were only one-third of Headquarters' estimated needs which approached 3 million per month.[63]

The satisfaction of such a large demand required powers which the Special Commission did not possess. To maximize production it was necessary to regulate and organize as well as to expedite. Armaments production was only the tip of the war-industries iceberg. Below it lay the metallurgical and fuel industries as well as the transport system. Light industry also produced for the army and required an effectively operating transport system, adequate supplies of raw materials, machinery and labour power. Thus, war-industrial production encompassed the entire commercial-industrial network. The separate and competitive demands of each unit in this network set in motion vicious cycles which the Special Commission could

do little to stop. The trains which hauled
supplies to the front could not at the same time
transport coal and iron to the factories producing
those supplies. The steel capacity that was used
for armaments could not go into the production of
rolling stock.

The Vankov Organization, named after Major-
General S.N. Vankov who founded and directed it,
represented an improvement over the Special
Commission as well as the regular military organs
and policies in two respects. First, it contracted
only with those enterprises which had not
previously produced for the GAU. Second, it was
involved at every stage of the productive process,
carrying out experiments, purchasing steel,
obtaining first priority in transport deliveries,
and recruiting skilled workers and soldiers.

The organization concentrated on the production
of the single piece 3-inch high explosive shell,
introduced into Russia by a French delegation in
January 1915. Between 1915 and 1918, it managed to
distribute orders for over 12 million of the
shells or 44 per cent of the total produced in
Russia during the war.[64] The organization's
staff was justly proud of its achievements, and in
1918 for the record and perhaps the eyes of the
new regime, produced a handsome volume which
documented and illustrated its methods. The
trouble was that the vicious cycles at work in the
Russian economy meant that Vankov's thoroughness
in providing his contractors with the means of
production detracted from other endeavours equally
as vital to the war effort. The high profits
accrued by Vankov's contractors (47 per cent
guaranteed gross) contributed to the fall in
metallurgical production in general and rails in
particular during 1915-16.[65] Either each article
required its Vankov, or the whole system had to be
coordinated.

Regardless of the organization's performance, it
had no effect whatsoever on the Galician campaign.
Created in March 1915, it was able to conclude the
first contracts only on 22 May.[66] In the meantime, the Russian armies which had entered the
Carpathians with hopes of driving on toward Vienna
and Budapest, were in retreat, and the issue of
supplies had become politicized. For this no one
was more responsible than the president of the

State Duma, M.V. Rodzianko.

Since the beginning of the war, Rodzianko had made several trips to the front. In mid-April he visited the Galician front observing the operations of the Third, Eighth and Ninth Armies. 'All evidence led to one main conclusion: that in the armies there was a grave shortage of ammunition', he later wrote.[67] Rodzianko managed to be in L'vov, the capital of Eastern Galicia, coincidental with the Tsar's triumphant visit. To the ever vigilant Minister of Internal Affairs, N.A. Maklakov, it was more than a coincidence. In a letter to the Tsar dated 27 April, Maklakov warned of the intrigues carried out by the 'pompous and stupid' Rodzianko.[68] In point of fact, Rodzianko did not intrigue against the Tsar, but did disapprove of the Tsar's presence in L'vov because as he purportedly told his Sovereign,

the soil on which the Russian monarch has once set foot cannot be lightly surrendered: torrents of blood will be shed upon it, but nevertheless we shall not be able to hold it.[69]

Rodzianko's pessimistic prediction, which turned out to be all too accurate, was based on his knowledge that Russia's advance against the Austrians had been halted. Within days of the Tsar's departure, a combined Austrian-German force under General Mackensen launched an attack along the southern wing of the Galician front with massive heavy artillery. 'The Russian artillery', noted Bernard Pares, an observer, 'was practically silent.'[70] No wonder when, against over 200 heavy German guns, the Third Army possessed two 4.2-inch and two 6-inch howitzers.[71] No wonder that the Third Army was in a shambles, or as its general, Radko-Dmitriev, put it in his report of 27 April, 'has literally lost all its blood'.[72]

Such was the state of affairs when Rodzianko, on his way back to Petrograd, stopped at Headquarters. Taking note of the atmosphere of depression, he felt obliged to reproach the Tsar's uncle, Nikolai Nikolaevich, quite sharply ('*ne do tseremonii*'). He accused the Supreme Commander of sacrificing the men in vain and demanded that he obtain from the GAU an exact account of what was being produced and what was ready. The reply of

Nikolai Nikolaevich revealed the complete break-
down in relations between Headquarters and the
Ministry of War. 'I cannot obtain anything from
the artillery department', he complained. 'My
position is very difficult.' V.P. Litvinov-
Falinskii, appointed to carry out an investigation
of factories working for defence, had been
inexplicably dismissed. Sukhomlinov was too much
in the Tsar's favour for the Supreme Commander's
influence over his nephew to have any effect.[73]

At this juncture, Rodzianko proposed what he
refers to as 'my old scheme', though there is no
previous mention of it anywhere - a committee
consisting of government officials, members of the
two legislative chambers, and representatives from
various industries 'with wide powers to deal with
all matters of war supplies'.[74] The project for
a Special Council was launched. Pursuant of the
project, Rodzianko held discussions in Petrograd
with Litvinov-Falinskii and his fellow Octobrist
Duma deputies, N.V. Savich, I.I. Dmitriukov and
A.D. Protopopov. In the second week in May, he
arrived back at Headquarters accompanied by
Litvinov-Falinskii and two Petrograd financier-
industrialists, A.I. Vyshnegradskii and A.I.
Putilov. Aleksandr Guchkov, 'inacceptable (sic)
but inevitable', was also invited.[75] After
Rodzianko reiterated his plan to the Tsar, the
outlines of the organization were immediately
drafted. For once the bureaucracy moved with
swiftness. On 13 May, Sukhomlinov received a cable
from Headquarters summarizing the plan and urging
him to proceed with its execution as soon as possible. On the very next day, Sukhomlinov presided
over the first session of what became officially
known as the Special Council for the Coordination
of Measures to Guarantee the Supply of Munitions
and Other Material to the Army. (The Special
Council went through a number of name changes
reflecting the expansion of its tasks. The one
cited here was that which appeared in the statutes
of 7 June. It will be abbreviated hereafter as
Special Council.)

Until 7 June when the statutes received the
Tsar's confirmation, the Special Council existed
as a union of convenience between government
officials - the majority of whom were from the
Ministry of War - four Octobrist deputies from the

Duma, and a 'commercial-industrial circle'.[76]
A.L. Sidorov has evaluated the presence of the
latter group in the following terms:

> This was a big victory for the bourgeoisie,
> obtaining the right to decide on an equal basis
> all questions of material provision for the
> army. Although in the council the generals
> constituted a majority, they already realized
> that without close cooperation with the bour-
> geoisie the situation with respect to shells and
> arms would not quickly improve.[77]

But not just any 'bourgeois' could sit in the
Special Council. So long as Sukhomlinov was chairman, only the traditional Petrograd-based munitions suppliers and their financial backers had
this privilege. At the four sessions held between
27 May and 13 June, eleven were present including
the chairmen of the boards of three large Petrograd commercial banks and the director of a
fourth. The Russian Company of Artillery Factories
(RAOAZ) was represented by four of the eleven,
and the Russian Company for the Production of
Shells, the Kolomna, Sormovo and Putilov Works,
the Petrograd Metal Works, and the 'Phoenix'
Machine Works by two each.[78]
   Sukhomlinov's faith in the resources of this
group was boundless. He assured the Council of
Ministers on 29 May that

> the banks have given directions to the factories
> which combined with them, can fulfil our orders.
> Putilov organized such a group which will
> prepare shells in full.[79]

The Special Council's session on 6 June, the last
over which Sukhomlinov presided, approved
Putilov's offer to be responsible for the
production of 3 million 3-inch shells, despite the
fact that the cost per shell would be 20 per cent
above what the Vankov Organization was paying its
contractors. With the assistance of the Russian-
Asian and Petrograd International Banks, the
Putilov Works was to arrange for the distribution
of orders worth 113.25 million rubles among nine
firms. Also approved at the session was an order
for 700 3-inch cannons to the Putilov Works at a

cost of 30 per cent per cannon above what had previously been paid.[80]

These orders reflect an extension of the policy of September 1914. The only real difference was the fact that now representatives of the firms responsible for the fulfilment of orders participated in their distribution. Such an arrangement constituted a form of state-monopoly capitalism, but by its very exclusiveness it provoked a hostile reaction by other industrialists. At the same time that Sukhomlinov was pleading before the Council of Ministers for the adoption of Putilov's proposal and the statutes of the Special Council, events elsewhere in the capital threatened to overtake both. Proclaiming its responsibility to mobilize all Russian industry for the war effort, the ninth congress of the Association of Industry and Trade struck on the idea of a network of War-Industries Committees.

3 Moscow to the Rescue: the Formation of the War-Industries Committees

The shortage of munitions was not the only factor in the Russian army's retreat from Galicia and Poland in the spring and summer of 1915. Poor handling by General Headquarters of what shells were available, tactical blunders, the lack of coordination between artillery and infantry, and problems with the distribution of other supplies – boots, packs, tents, food, etc. – all played their part.[1] Nor was the Russian army the only one to suffer from inadequate quantities of shells. Between November 1914 and February 1915, the British army received between one-fifth and one-tenth of what it required in the way of shells and by the end of May 1915, of the 5.5 million shell casings due for delivery, only 1.9 million had arrived.[2]

In Britain, as in Russia, a political crisis erupted in May 1915 over the shortage of munitions. The crisis in Britain was resolved by the transformation of the Asquith government into a coalition cabinet containing Liberals, Conservatives and a member of the Labour Party; the formation of a Ministry of Munitions under Lloyd George; and the subsequent reorganization of the supply system. In Russia, two attempts at a solution emerged. The formation of the Special Council was one. The other was embodied in a resolution adopted by the Association of Industry and Trade's ninth congress.

The resolution was by no means a unanimous decision on the part of Russian industrialists. Those who had ensconced themselves in the Special Council were sufficiently well-placed to ignore it. Other Petrograd industrialists attempted to minimize its significance. Even the leaders of the

Association of Industry and Trade were ambivalent
about its implications. But as the retreat from
the German onslaught continued, with the loss of
Libau and Galicia, then the Polish fortresses, and
by August the evacuation of Warsaw and Riga, the
Russian government and those favoured by it were,
like the army, thrown back onto the defensive. It
was in such circumstances that the War-Industries
Committees came to represent a force to be
reckoned with in the political and economic life
of the country.

3.1 RIABUSHINSKII AND THE MOBILIZATION OF RUSSIAN INDUSTRIALISTS

The peak of the retreat from Galicia coincided
with the ninth congress of the Association of
Industry and Trade. The fact that the congress
directed itself to the immediate crisis rather
than proceeding with its agenda constituted an
important change in the relationship of the
industrial bourgeoisie to the war effort. It was a
change that would have been difficult to predict
before the congress. After its January 1915
proposal (see pp. 33-4), the Association slipped
back into its normal routine of narrow professional interests and quietism. Less than two
months before the congress, its journal was
assuring the public that

> Russia is not going through any economic crisis.
> The effort exerted by the government in the
> military struggle is reflected quite insignifi-
> cantly in the economic sphere of the country.
> Demand for all goods is great and the productive
> capacity of factories can at any moment be
> maximized.[3]

There are two points to note about this statement.
First, the Association failed to make any connection between 'the effort exerted by the government
in the military struggle' and 'the economic
sphere'. Second, the assertion that 'demand...is
great' was juxtaposed with 'productive capacity...
can...be maximized', implying that for the time
being there was no need to maximize production.
But what if, even in the Association's terms, an

economic crisis were to occur? The editorial
continues: 'The main condition for the correct
resolution of the most difficult economic questions is calmness.' Nowhere was there a call for
or mention of mobilization, of sacrifices, of the
fact that this was not a war which only soldiers
fought. Clearly, 'organized industry' in the
period before the Galician retreat had failed to
recognize the necessity of *Umstellung*, of the
mobilization of industrial resources to fulfil the
needs of war.

In fact, the programme for the ninth congress
struck many by its resemblance to pre-war agendas.
There were to be four major addresses - I.P.
Bragin on state monopolies, V.V. Zhukovskii on the
raising of productivity in Russian industry, V.P.
Litvinov-Falinskii on industry and the war, and
N.S. Avdakov, the chairman, on the establishment
of a supreme economic commission to determine
post-war policies. On the day the congress
convened, *Russkie Vedomosti* (Russian Gazette),
identifying itself with Kadet Party thought,
published an editorial which severely criticized
the retention of this programme in the light of
the desperate situation at the front. It predicted
none the less that

> As always, questions of our economic development
> will be resolved by the industrialists in
> accordance with the needs of the whole country
> only to the extent that this coincides with the
> interest of the large enterprises. In those
> cases when the interests of the country come
> into conflict with the class needs of the
> factory owners, preference will be given to the
> latter.[4]

S.O. Zagorskii, the economic correspondent for
Den' (The Day), chided the Association for its
indifference to the plight of the country. 'All
this time what has industry done? Not one step has
been taken toward putting its financial resources
at the disposal of the nation', he complained.[5]

In Moscow, especially, a ground swell of opinion
arose against the Association's efforts to promote
business as usual. 'We can regard the forthcoming
congress as nothing but a reflection of specific
narrow professional interests', declared the organ

of small and medium-sized businesses. 'It does not represent the public, nor does it unite all-Russian commerce and industry.' The paper viewed Avdakov's proposed commission as an attempt to prevent the erosion of the syndicates' power as well as that of 'the staff of the all-Russian syndicates, the Association'.[6] Finally, on the day the congress opened, *Utro Rossii* questioned its priorities in the following terms:

> The programme of the commercial-industrial congress was outlined and formulated long before the present time. Then it was called 'Industry and the War'. Now, we must say 'The War and Industry'. Then, a discussion was proposed on the interests and tasks of Russian commerce and industry in the future, after the war. Now, interest in the future must be cast aside before the overwhelming force of present concerns.[7]

What those concerns were and what industrialists could do about them had already been discussed in the pages of *Utro Rossii* by several prominent professors, engineers and industrialists. Their contributions appeared under the banner headline 'All for the War', which was soon to become the slogan of the War-Industries Committees.[8]

Nevertheless, the congress did not alter its agenda. On the opening day, nearly 500 delegates plus prominent members of the Duma and State Council heard Avdakov deliver his introductory remarks and Litvinov-Falinskii his address on industry and the war. Litvinov-Falinskii warned that Russia must avoid reassuming its role as an economic colony of Germany after the conclusion of hostilities. The war, in his view, provided an opportunity for Russian industry to fill the vacuum created by the severance of economic ties with Germany. However, the liquidation of German property and the boycott of German goods, steps ardently advocated by the right-wing press since the earliest days of the war, were not mentioned.[9] Instead, he proposed revisions in the corporate laws to attract foreign capital (presumably with the exception of that originating in Germany), and the adoption of the Kharitonov Commission's recommendations on state orders. Much in line with Zhukovskii's projected report,

published beforehand, the purport of Litvinov-
Falinskii's address was not what industry could
contribute to the war effort but rather the
reverse.[10]

Bragin read his speech on the second day of the
congress, and there followed some desultory
discussion. Then, Pavel Pavlovich Riabushinskii,
recently elected as vice-chairman of the Moscow
Exchange Committee, took the floor. The speech
which he delivered led to the scrapping of the
congress' agenda and the tabling of a discussion
of how to provide the means to defeat the enemy.

Riabushinskii had come directly from the front
to address the 27 May session. With a voice
described by one reporter as dry and hard, raised
now to a cry and then lowered to a whisper, he
commanded the attention of his audience.[11] He
contrasted the firm resolve of the soldiers with
the insufficient work done in the rear. He told
the delegates that it was their responsibility,
individually, to do everything possible to
minimize the sacrifice of lives. They had to adopt
the same determination which the Germans exhibited. Employing the logic of social Darwinism, he
predicted that 'those who move forward sooner will
live longer. The unsteady will fall down.' 'You
know yourselves', he continued,

that every factory in Germany previously
satisfying demands of the whole world is now
directed to the sole unshakeable purpose of
contributing to the success of their army.

But what exactly were the delegates to do?

We can no longer occupy ourselves with everyday
matters. Each factory, each mill, all of us must
only think of that which can overcome this enemy
force....We need soldiers with guns; we need
machine guns; we need an infinite amount of
barbed wire, and only then, with the strong
resolve which the Russian army possesses, will
we overcome the enemy....It is a small matter,
gentlemen, to proclaim the necessity of attracting knowledgeable people to create a committee.
God knows we have created committees quite
often, but usually they have turned out to be
lifeless or stillborn. Therefore, we need to

think...about the possibility of attracting worthy and experienced as well as knowledgeable people from the government, society, industry, etc. We need leaders and such leaders we have.[12]

The contrast between this position and that outlined by Litvinov-Falinskii is striking. Yet, it would be a mistake to categorize it merely in terms of a war as opportunity/war as sacrifice dichotomy. Basic to Riabushinskii's demand for sacrifices was his understanding of the immense opportunities which not only the munitions shortage but the supply crisis in general provided Russian industrialists. He was at once the advocate and practitioner of retooling industry and redirecting investments to meet the needs of war. 'One must not laugh at the Germans for growing potatoes in their flower beds and hot houses', he told the Moscow Exchange Committee in April 1915.[13] During the war the Riabushinskii family invested in two of Russia's export industries - lumber and gold - and two defence industries - coal and automobile production.[14]

If Riabushinskii linked the munitions shortage to all supplies required by the army, and the satisfaction of the army's requirements with economic development, he also recognized that economic development was impossible without political power. Thus, industrialists were to organize themselves not simply to obtain more lucrative contracts as Litvinov-Falinskii implied. This could only be temporarily profitable but in the long run disastrous, for they would be obtaining these orders from a government whose incompetence made the fulfilment of orders difficult and in fact threatened the very survival of Russian industry. The government itself had to be reorganized to put Russia's war economy on a sound footing, and in the absence of any other group, Russia's patriotic industrialists and technical personnel would reorganize it.

Taking the podium after Riabushinskii, V.V. Zhukovskii attempted to minimize the effect of his speech. Zhukovskii moved that Riabushinskii's proposals be incorporated into the resolutions based on Litvinov-Falinskii's report, which had been adopted the previous evening. However, the delegates, roused by Riabushinskii's alarm, refused to be pacified. M.M. Fedorov, board chairman of the Azov-Don Bank and an original member of the Progressist Party's central committee, was the first to challenge Zhukovskii's recommendation. Avdakov interjected that Rodzianko would discuss Riabushinskii's proposals at the 28 May session, but Iu. I. Poplavskii, vice-chairman of the Society of Factory and Mill Owners of the Moscow Region and also a Progressist, asserted that the congress would have nothing of it.[15] The Muscovites had clearly won the day.

After a short recess, a special commission of 35 was designated to draw up resolutions and the remaining speeches were cancelled. Zhukovskii admonished the delegates to formulate resolutions which would have a 'calm business-like character' because 'to help our glorious army and head it to victory we do not need animosity or meetings and protests but serious practical steps'.[16] This time, the delegates heeded his advice. The special commission drafted a three-part resolution which the next session of the congress unanimously approved. The resolution, the founding charter of the War-Industries Committees, deserves to be quoted in full:

Recognizing the inevitability of the application of all productive forces to the organization of work for defence, in view of the extreme necessity for a better organization of the rear, the Association of Industry and Trade, suspending discussion of the proposed program of the congress until a more propitious time, unanimously resolves:
(1) to organize all the unutilized power of Russian industry for the satisfaction of the needs of the state's defence;
(a) to instruct all commercial-industrial organizations to form regional committees for the unification of local commerce and industry with the aim of defining the capabilities of enterprises for producing all that is necessary for the army and navy, and, coordinating the activities of factories and mills, to draft a plan for the urgent fulfilment of present work as well as to define the requirements in raw materials, fuel, transport, and labour force;

(b) for the coordination of all work of various regions and groups, and equally for the coordination of this work with the activities of the High Governmental Institutions, the congress resolves to establish in Petrograd a Central War-Industries Committee, assigning the organization of the Committee to the Council of the Association such that representatives of science and technology, from various commercial-industrial organizations, from the railway administrations and the All-Russian Unions of Zemstvos and Towns may participate;

(2) to propose to the Central War-Industries Committee that it turn its attention to questions of communication and transportation, to ensure industry the necessary means for which the congress considers advantageous the division of the committee into sections. The elaboration of a plan of general activity, to be communicated to local organizations, must be the first order of business of the Committee as far as organizational work is concerned;

(3) to cover organizational expenses of the War-Industries Committee by preliminarily assigning 25,000 rubles, and directing the Council to work out a budget of expenses and ways of meeting them.

The activity of the War-Industries Committee must be directed towards the satisfaction of the requirements of the army's Supreme Command. The congress expresses its confidence that all Russian industry will whole-heartedly give itself to this matter and find within itself the strength to deal with the great historic tasks. The best and most capable people must be attracted to the business of administration to fulfil them.

In recognizing the necessity to carry the war to a victorious conclusion, of not being stopped by any sacrifice, by the length of the struggle, and in the firm conviction that our glorious army under the leadership of the Supreme Commander in close cooperation with the nation will fulfil this great historic task, the Association of Industry and Trade calls on the entire country, on employees and workers, to establish a plan of work for industry in a calm and systematic manner with the consciousness of

full responsibility which this historic moment
places on Russian industry. The War Committee of
Russian Industry, using the guidelines of
previous experience, must devote itself to the
realization of the tasks placed before it,
making good all the deficiencies of supplies.
The congress expresses confidence that Russian
industry will carry out its duty to the country
and will deserve the praise of the mighty Ruler
of the Russian Land on Whom all fix their gaze.
The congress instructs the Committee at this
moment to address His Imperial Majesty through a
special delegation with an expression of loyalty
from Russian industry and commerce.[17]

Despite the obsequiousness of its concluding para-
graphs, the resolution was a victory for the
Riabushinskii initiative. It represented a bold
attempt by Russia's 'patriotic' industrialists at
the head of whom stood Moscow's manufacturer-
Progressists, to project onto an all-Russian scale
a plan for victory in the war and through it the
future industrial development of the country. In
this sense, and not this sense alone, the War-
Industries Committees emerged more as an
aspiration than an application of already existing
resources.

The aspirations of these industrialists were
reflected in the name of the organization they
created. The term 'War-Industries Committees'
(*voenno-promyshlennye komitety*) appears to have
been borrowed from the German organization,
Kriegsausschuss der Deutschen Industrie, formed
during the first days of the war by a merger of
Germany's two major industrialist federations.[18]
(The Russian translation was *voenno-promyshlennyi
komitet*, or less often *voennyi komitet nemetskoi
promyshlennosti*. Hereafter, the War-Industries
Committees will be referred to as VPKs, the
abbreviation of their Russian name.) It was widely
believed in Russia that the *Kriegsausschuss*
wielded enormous power over the distribution of
state orders, the regulation of materials and
goods, the labour market, the technical adaptation
of enterprises to war production, and the control
of imports and exports.[19] Given German
industry's reputation for efficiency, there was
every reason for its Russian namesake to seek to

emulate it.

These industrialists, however, knew only too well that while the German government welcomed the initiative taken by German industry, the Russian government, traditionally suspicious of autonomous 'social' movements, would almost certainly not look kindly on the VPKs. They and their allies within the Duma's recently formed Progressive Bloc therefore developed the idea of a Russian ministry of munitions on the order of that created in Britain under Lloyd George. If political pressure could elevate a representative of 'society' or, preferably, the committees themselves to a ministerial position, they were prepared to exert that pressure. They would seek to transform the organization into a network of district committees, of which the one established in Leeds on 1/13 May was prototypical.[20]

But, as Lloyd George later wrote, 'the Ministry of Munitions was from first to last a business-man organization'.[21] Before the committees could achieve the political power necessary to rule over the economy, they had to win the allegiance of Russian businessmen at large. This, then, was the first task which confronted the delegates to the ninth congress after they dispersed to the provinces.

3.2 THE DISTRICT AND LOCAL WAR-INDUSTRIES COMMITTEES

The response to the Association's resolution by exchange committees, zemstvo and town councils, and technical and industrial organizations was tremendous. In large industrial centres as well as small administrative and commercial towns, in those areas near the front as well as those far removed from it, even in Siberia and the Far East townspeople gathered to discuss the resolution, and as it instructed, to organize committees.

On 30 May 1915, the Association of South Russian Mineowners established a VPK which was supposed to encompass Khar'kov, Ekaterinoslav, Kursk, Orel and Voronezh provinces. It sent telegrams of solidarity to Rodzianko and Avdakov.[22] Within a few days, industrialists in Moscow and Odessa organized committees. Then came Kiev, Riga,

Saratov and Simbirsk. At least a dozen committees were created between 15 and 21 June. By 10 July, 67 existed, and 2 weeks later, when the first all-Russian congress convened, delegates from 78 VPKs plus 7 affiliated bodies attended.[23]

Thereafter, the organization's growth was primarily due to the assistance of already established committees. By November 1915, the central committee reported 31 district and 81 local VPKs. And when the Second All-Russian Congress met in February 1916, there were 34 district and 192 local committees.[24]

In Vitebsk, Moscow, Kursk, Nizhnii Novgorod, Mariupol', Novorossiisk, Ekaterinodar, Samara, Perm, Ekaterinburg and Cheliabinsk, the local exchange committees which had represented their respective business communities for decades supplied the initiative.[25] In Vilna, Smolensk, Kherson, Saratov, Penza, Tiflis, Kazan' and Krasnoiarsk, city dumas fulfilled this function.[26] In some of the larger industrial and commercial centres such as Petrograd, Khar'kov, Ivanovo-Voznesensk, Odessa and Baku, regionally based industrialist organizations repeated on a smaller scale that which had occurred at the ninth congress by forming the nuclei of district VPKs. And in Minsk, Poltava, Rybinsk, Riazan' and Kremenchug, the local zemstvo board (*uprava*) or branch of the All-Russian Zemstvo Union and/or Union of Towns spawned committees.[27]

Apart from the enfranchised local population, representatives of other groups occasionally made bids to form VPKs. In Krasnoiarsk, the governor-general refused to allow the committee organized by the city duma to meet, not because he opposed the movement but because he had organized one consisting of military and administrative personnel.[28] In Orenburg, by contrast, a 'citizens' committee' was formed on 16 July in which several workers were included. Three days later, the committee sponsored a meeting attended by 500 to 600 people, and sent to the first all-Russian congress of VPKs the engineer, P.A. Kobozev.[29] On his return from Petrograd, Kobozev addressed a crowd estimated at 5000, but immediately thereafter was arrested along with a number of workers.[30] The city and zemstvo officials seem to have prevented further participation by workers in the committee, but did not terminate Kobozev's association. At the second congress of VPKs in February 1916 he turned up to denounce the 'deep penetration' of the right into the Orenburg district VPK, as exemplified by the election of the Turgai provincial vice-governor to the vice-chairmanship of the committee.[31]

Until the central committee launched its campaign in August to attract representatives of the working class, they did not appear in the VPKs, with Orenburg's exception. The industrial bourgeoisie, on the other hand, constituted the essential social element in the VPKs from the outset. Its domination of the organization was codified in the instructions drawn up for local committees by the Moscow VPK. Committees would be approved, it was stipulated, 'if they represented organizations of industrialists working for defence and *included* representatives of other elements'.[32] Table 3.1, adapted from a survey published by the central committee in 1916, shows that representatives of commercial-industrial organizations comprised the largest group: 35.6 per cent in district committees and 30.1 per cent in local committees.

The table actually underestimates the number of those involved in industry and commerce, for it measures the group affiliation of individuals rather than their occupation. Many of those listed as city and zemstvo officials were at the same time industrialists and merchants. For example, the 160-member Moscow city duma contained 72 merchants, industrial employers, and directors of large companies in late 1916.[33] D.V. Sirotkin, the chairman of the Nizhnii Novgorod district VPK, was mayor of the city, but he was also the chairman of the exchange committee, board chairman of a number of grain shipping companies, and a board member of the Sormovo metalworks plant located in the city environs.[34] The fact that the percentage of 'others' in local committees was twice that of district committees may be correlated with the lower percentage of representatives from commercial-industrial organizations. Though these organizations reached down to the small towns, they were not as extensive as in the larger industrial and commercial centres.

TABLE 3.1   *Social composition of district and local committees, 1916*

Groups	20 district VPKs		98 Local VPKs		Total	
	No.	%	No.	%	No.	%
1 City/zemstvo officials	177	16.9	574	17.1	751	17.1
2 Government officials	112	10.7	358	10.7	470	10.7
3 Tech./scientific orgs	216	20.6	497	14.8	713	16.2
4 Comm.-industrial orgs	372	35.6	1006	30.1	1378	31.4
5 Members of cooperatives	17	1.6	101	3.0	118	2.7
6 Workers	35	3.4	72	2.2	107	2.4
7 Others*	117	11.2	739	22.1	856	19.5
Total	1046	100	3347	100	4393	100

* Includes individuals of free professions, unaffiliated industrialists and businessmen, and those specially invited by the committees.

Source: *Predstavitel'stvo obshchestvennykh grupp v VPKakh.* This survey is by no means complete, lacking information from 14 district and 94 local committees. It can, to some extent, be supplemented with the *Lichnyi sostav voennopromyshlennykh komitetov* (Petrograd, 1915). A rough approximation of the total number of VPK members would be 9000. This does not include office staff and other employees.

Occasionally, members of the committees were incautious enough to openly admit the class nature of the organization. At the Moscow VPK's session on 28 February 1917, in the thick of the February Revolution, one member, seeking to define the committee's relation to the events in Petrograd, stated:

The Moscow VPK represents the interests and needs of industry. It is an organization with a class nature. It unites industrialists for the satisfaction of their immediate requirements.[35]

Another member, writing in the Moscow committee's *Izvestiia* several months earlier, referred to the organization as one which 'unites all sectors of Russia's progressive bourgeoisie'.[36]

But such characterizations were rare. Officially, the VPKs were a 'public' (*obshchestvennaia*) organization, sharing this distinction with the Union of Towns, the Zemstvo Union, and their jointly staffed Main Committee for Supply of the Army (Zemgor). According to A.I. Guchkov, speaking in May 1917, 'the committees are outside of any class (*vneklassnye*)'.[37] A central committee English-language publication of 1918 described the committee as 'the most democratic public organization under the old regime'.[38]

What lent some credibility to these statements was the fact that other than industrialists, another distinct group was well represented and played a significant role in the organization. It consisted of engineering-technical personnel and other professional people whose technical skills were required as much as the industrialists' entrepreneurial and organizational experience in the mobilization of Russian industry.[39]

From their inception, the committees appealed to scientists and engineers to throw in their lot with the movement, and in this respect, they were rather successful. It was not so much that Russia's technical intelligentsia lacked its own organizations, but that such organizations tended to isolate their members from the broad political and social currents, making 'orphans' of their members.[40] The Imperial Russian Technical Society's Commission on Industry in Connection with the War was a case in point. Organized in September 1914, it had drawn up a number of insightful reports on machine construction, shell production and fuel supply.[41] But these reports were of use to nobody because nobody bothered to use them. Once an organization had been created for the explicit purpose of applying 'all the

unutilized power of Russian industry for the
satisfaction of the needs of the state's defence',
the Commission decided to dissolve itself and
offer the services of its members to the VPKs.[42]
In this way, P.I. Pal'chinskii, M.N. Novorusskii,
V.I. Kovalevskii and D.S. Zernov entered the
central committee.

For the most part, though, scientists and engineers did not need their respective organizations
to advise them to join the committees. Their sense
of patriotic duty, hitherto unexploited by the
government, and their appreciation of the
importance of technology in the war effort were
sufficient to stimulate them. Hence, the Moscow
committee accommodated Professors A.E. Chichibabin, N.E. Zhukovskii and V.I. Grinevetskii,
pioneers in their respective fields of chemistry,
physics and thermodynamics. In Baku, the oil
producing centre of the empire, the district VPK
included the director of the Baku Technical
School, the vice-chairman of the Imperial Russian
Technical Society's local branch, and several
mining engineers and technical directors employed
by the giant oil firms. In Grozny, second to Baku
in oil production, the VPK consisted *inter alia* of
the chairman of the local branch of the Imperial
Russian Technical society and representatives from
the city's 'Neft'' and Nobel plants, as well as
locally owned oil enterprises.[43] The Kiev
committee was perhaps the most successful in
mobilizing the local technical intelligentsia. Of
a total of 103 members in 1916, 24 were professors
- 15 of whom were connected with the Kiev
Polytechnical Institute - and 12 were engineers.[44]

Scientific and technical personnel rarely occupied positions of leadership within the VPKs. They
only furnished half as many chairmen and vice-
chairmen as city and zemstvo officials, although
they were nearly as numerous in the organization.[45] Generally apolitical, they served the
committees - as many of them would serve the
soviet state after 1917 - in the important but
largely unheralded sphere of technological
research and enterprise management.

The hegemony of the industrial bourgeoisie
within the VPKs was both reflected in, and ensured
by, the election of provincial and local business

magnates to leading positions. The Baku committee
was directed by the oil entrepreneur and chairman
of the Association of Baku Oil Producers, A.O.
Gukasov. The Kiev committee was chaired by the
sugar beet millionaire, M.I. Tereshchenko, who
guaranteed it one of his millions in capital.[46]
In Odessa, the VPK contained thirteen members of
the Odessa Society of Factory and Mill Owners.
S.I. Sokolovskii, the chairman of the Society and
according to one account the 'Figaro of Odessa
industrial organizations', served as vice-chairman
of the VPK.[47] In Rostov-on-Don, the Paramonov
brothers, Nikolai and Petr, dominated the industrial life of the city. The family, known as the
'Croesus of Rostov', owned several anthracite
mines, the largest milling complex in the region,
and a local newspaper. Nikolai was chairman of the
district VPK while Petr was one of three vice-
chairmen.[48]
    As will be demonstrated below, there were huge
differences in the degree to which these
committees managed to mobilize industry within
their respective domains. Some district
committees, after the initial burst of organizational activity, lapsed into the performance of
registrational/archival functions; some local
committees, having established close connections
with other public organizations, extended their
activities to include food distribution, the
collection of medical supplies and other functions
only distantly related to the supposed purpose of
the movement. But the fate of the organization as
a whole really hinged on the orientatation of
activities of the Central and Moscow VPKs.

3.3 THE CENTRAL AND MOSCOW WAR-INDUSTRIES
     COMMITTEES

In the exultation which followed the Association
of Industry and Trade's ninth congress, few
noticed what had become of the organization
spawned by it. *Birzhevye Vedomosti* congratulated
the Association for its conscientiousness and
referred to its members as the 'true sons of
Russia'. Zagorskii, who only two days previously
had attacked industrialists for their indifference
to Russia's fate, rejoiced that *'fata volentem*

ducunt,   nolentem   trahunt!'   ('Fate   leads   the
willing   and   the   unwilling   alike!')[49]   In   fact,
the   unwilling - the   Association's   executive   body -
after   initial   reticence,   took   fate   into   its   hands
and   sacrificed   little.   Section   1(b)   of   the   resolution   drafted   by   the   congress'   special   commission
invested   the   council   of   the   Association   with   the
organization   of   the   Central   War-Industries
Committee   (hereafter   TsVPK).   The   day   after   the
congress   had   closed,   Avdakov,   as   chairman,   urged
the   council   to   take   exclusive   control   of   the
TsVPK's   organization   and   'in   the   interests   of
efficiency'   to   limit   the   number   of   members.[50]
Another   preparatory   meeting   followed   before
Avdakov   presided   over   the   first   session   of   the
TsVPK,   on   4   June,   at   which   his   dictum   was   apparently   realized.   The   council   decided   that   it   was   to
constitute   the   core   of   the   TsVPK,   coopting
representatives   of   district   committees   as   they
came   into   existence.   But   since   the   number   of
district   committees   was   so   large,   the   TsVPK,   on
Zhukovskii's   recommendation,   formed   a   Presidium
drawn   from   the   Association's   council   and   headed   by
Avdakov.[51]   After   16   June,   the   TsVPK   which   had
been   meeting   on   a   daily   basis,   convened   only   fort-
nightly,   and   eventually,   once   a   month.   Between   4
August   1915   and   12   September   1916   it   held   only   25
plenary   sessions.   From   September   until   the
February   Revolution,   the   TsVPK   did   not   meet   as
such,   but   hosted   three   extraordinary   gatherings
with   district   representatives.[52]
   The   Presidium,   later   renamed   the   Bureau,   had
ultimate   responsibility   for   all   the   central
committee's   sections.   It   determined   the   allocation
of   moneys,   state   contracts   and   materials   obtained
from   the   state.   It   was,   in   addition,   the   final
adjudicator   of   all   disputes   arising   within   a
particular   committee,   between   two   committees,   or
involving   a   committee   and   its   contracting
partners.
   The   real   work   of   industrial   mobilization   fell   to
the   TsVPK's   sections   which   originally   numbered
thirteen: mechanical,   chemical,   clothing,   food,
medical,   inventions,   transport,   fuel,   labour
supply,   metallurgical,   electrical   power,   legal   and
financial.   Managers,   engineers   and   technicians
congregated   in   these   sections.   D.S.   Zernov,   a   past
director   of   the   Petersburg   Technical   Institute,

replaced Zhukovskii as chairman of the all-important mechanical section at its session on 16 June. Directors of some of the largest metalworks enterprises in the country - Breyer of the Putilov Works, K.P. Fedorov of Petrograd Metal, and N.P. Koksharov of the Briansk Railrolling, Smelting and Ironworks Company - attended this session. By August, the section boasted 15 engineers, 9 technicians and 14 students.[53] The electrical power section was graced by a competent staff which included the future Soviet Commissar of Foreign Trade, L.B. Krasin. The director of the Siemens-Schuckert plant in Petrograd, Krasin took an active part in the VPKs' affairs. At one early session he boldly suggested that Russia import dynamos from Germany and Austria-Hungary because 'now is not the time to think of protectionism.... We need to think only of victory and for this all paths are permissible.'[54]

Many of the Association of Industry and Trade's executive staff, especially those with technical backgrounds, became chairmen of the TsVPK's sections. These were managers and engineers with a stake in industrial development. N.N. Iznar, the Association of Baku Oil Producers' representative in Petrograd, the director of the Southeastern Railroad, and a long-standing member of the Association of Industry and Trade's committee, was head of the military stores section. A.A. Bublikov, a Progressist deputy in the Fourth Duma and like Iznar a member of the Association's committee, served as chairman of the transport section. N.F. Fon-Ditmar, chairman of the Association of South Russian Mineowners, took charge of the fuel and labour supply sections.

They and others from the Association - N.N. Kutler, P.P. Kozakevich, Baron G. Kh. Maidel and V.V. Zhukovskii - provided a pool of administrative and technical talent on which the TsVPK heavily depended. Even when relations between the Association and the committees became strained to breaking point over the latter's labour policy, these cadres remained active in both organizations. As the post-1917 careers of several of them demonstrate, their political adaptability was not limited to internecine conflicts within the industrial bourgeoisie.[55]

In June 1915, it was to the TsVPK, chaired by

Avdakov, that they gave their services, thus making the new organization's apparatus virtually indistinguishable from that of the Association. Avdakov and Iznar comprised the delegation which the ninth congress had instructed to represent the TsVPK before the Tsar. Avdakov, Maidel and Zhukovskii were its representatives in the Special Council.[56] In this manner, the Association enjoyed a near monopoly of responsibility and publicity in the early days of the TsVPK.

The Muscovites, however, had great success on their home territory. They returned to Moscow to treat the city to a veritable orgy of meetings, both official and private. While Riabushinskii spoke to the Moscow Exchange Committee, Iu. I. Poplavskii addressed the Society of Factory and Mill Owners, M.M. Fedorov the Main Committee of the Zemstvo Union, and A.I. Konovalov a private meeting of industrialists. Out of these gatherings arose the skeleton of a Moscow VPK. The exchange committee and the society of factory owners promptly contributed 25,000 rubles each toward its creation. The Zemstvo Union resolved to organize a meeting with the Union of Towns, delegate two representatives each to the Central and Moscow VPKs, and call on local zemstvo boards to send representatives to the appropriate local VPKs.[57]

On 4 June, the exchange committee established a Provisional Commission for the Organization of the Moscow War-Industries Committee. A week later the Moscow VPK held its first session under Riabushinskii. The initial composition of the committee differed considerably from that of the TsVPK, and the differences are instructive. The TsVPK attracted managers and engineers from across the country, but failed to attract Petrograd's leading industrialists.[58] The Moscow committee contained 14 representatives from merchant and exchange organizations, 7 from textile groups, 6 representing mixed factory and mill owner organizations, 5 from academic institutions, and 1 representing the machine-tool industry. There were no representatives from the Association of Industry and Trade, although several participants in the Moscow VPK were members. In short, the Moscow VPK reflected its commercial industrial constituency.

Throughout June that constituency boiled with

energy. As in 1905, when strikes in the city
provoked the business community to organize itself
in the form of employers unions, political parties
and other representative bodies, the stimulus for
such activity in 1915 came from below. At the very
moment that Riabushinskii stood before the
delegates at the ninth congress, in Moscow,
elements of the working class and *Lumpenprol-
etariat* were destroying suspected German property
under the apparently indifferent if not benevolent
gaze of the prefect and his police. The *pogrom*,
resulting in damage which the acting British
Consul-General described as 'greater than during
the [1905] revolution',[59] lasted three days.
Another British eyewitness reported that the
destruction was by no means confined to German
property.[60] The cry for the mobilization of
industry, then, had more than one motivation.

In addition to the meetings already mentioned,
both the Zemstvo Union and the Union of Towns held
their third congresses on 5 June. These two all-
Russian public organizations were, in reality,
controlled from Moscow. Moscow served as the
location of the Main Committees of both bodies as
well as for all congresses. It furnished most of
the scientific and technical personnel on the Main
Committees as well as Prince G.E. L'vov and M.V.
Chelnokov, the High Commissioners. The salient
point about the third congresses is that they
differed considerably from their predecessors.
'Patriotic enthusiasm and most faithful servant
telegrams' characterized the July/September 1914
and February/March 1915 gatherings.[61] But those
of June 1915 served as a forum for the expression
of opposition to the government's supply policies
and support for the Riabushinskii initiative.
Prince L'vov at once grasped the significance of
the moment, proclaiming that 'Russia must become
one vast military organization, a hugh arsenal for
the armies'.[62] Both Unions resolved to invest in
their Main Committees the responsibility to
organize supplies to the army (in the form of the
already mentioned Zemgor) and appoint representatives to the VPKs. Prince L'vov and Chelnokov were
present for the first session of the Moscow VPK on
11 June and shortly thereafter became vice-
chairmen of the rapidly growing organization.

Yet another distinguishing characteristic of the

Moscow VPK was its close connection with the Progressist Party. Three of four on the committee's Presidium and approximately half of the section chairmen can be identified with the party. While the Kadets, meeting simultaneously with the Unions in Moscow, berated themselves for their reflexive adherence to the policy of unity with the government, the Progressists mobilized themselves for the imminent Duma session.[63] They resurrected the idea of a left of centre progressive majority in the Fourth Duma excluding the small contingent of Social Democrats and Trudoviks. Outflanking the Kadets on the left, the Progressists successfully courted the Moscow branch leaders, N.I. Astrov and F.I. Rodichev, and placed the Moscow VPK in the forefront of the movement for a 'responsible ministry', that is, one responsible to the Duma.[64]

The Moscow VPK was a second central committee in all but name. It was the only district committee with the right to contract directly with the central military departments. At first, it consisted of five major divisions: (1) armaments, which included chemicals, machines and shells; (2) uniforms, which encompassed cotton, linens, wool and footwear sections; (3) transport; (4) labour supply; (5) food. But the proliferation of the committee's concerns soon compelled it to subdivide to the point where its sections eventually outnumbered those of the TsVPK. A financial account filed by the committee's auditing section in August 1916 listed nineteen operative sections.[65]

All four sections dealing with textiles were directed by major textile owners themselves: S.N. Tret'iakov (linens), N.T. Kashtanov (silk and wool) and St.P. Riabushinskii (cotton). The machine section occasionally included representatives from the main heavy industrial enterprises of the region as, for example, when it considered the coordination of fuel and raw material deliveries.[66] The shell section served as a link between the Moscow War-Industry Company (1915), a shell producing concern created simultaneously with the Moscow VPK, and the Vankov Organization which supplied the section with orders.[67] Having thus attracted the flower of Moscow's commercial, industrial and technical communities, the committee expressed its confidence that it could

> take on not only the role of distributor among
> separate enterprises, but would itself take the
> economic initiative, independently giving orders
> ...organizing its own control, and directly
> delivering goods to the army.[68]

At the same time, new committees appeared in the
provinces requesting permission to enter the
Moscow committee's sphere. From the middle of June
through July, the newspapers carried daily information on the establishment of such VPKs:
Simbirsk and Tver on 11 June, Voronezh on the
15th, Minsk on the 18th, Rybinsk on the 26th,
Ivanovo-Voznesensk on the 30th. By August the
Moscow VPK encompassed the entire Moscow industrial region, the central agricultural region
(minus Penza province), the Belorussian provinces
of Minsk and Smolensk, and Simbirsk in the east,
or fifteen provinces in all with an area the size
of France.[69]

These developments in Moscow and their
repetition on a smaller scale in other parts of
the empire were bound to affect the composition
and orientation of the TsVPK. They confronted it
with the choice of acknowledging and making the
most of the popular response to the call to mobilize private industry, or becoming, as one journal
predicted, 'the one thousand and first commission
of the Association of Industry and Trade to
distribute state orders among its favourite
factories'.[70]

The Petrograd banking community tried to
persuade the TsVPK to consider the merits of the
second alternative. A.I. Putilov revealed in
February 1916 that in the early weeks of the
TsVPK's activities he and unnamed others
approached the committee with a proposal for the
establishment of a special section devoted to
large metalworks firms.[71] The TsVPK's journal
for the third session held on 9 June 1915 does
indeed refer to 'several Petrograd banks' express-
ing a readiness to participate in the creation of
'a[n unspecified] section'.[72] But the TsVPK,
perhaps anticipating the reaction of the Moscow
and other committees, did not respond. The
Petrograd financiers thereafter remained outside

and, on certain issues, hostile to the VPKs. Further evidence that events in Moscow were influencing the TsVPK was the deterioration of the position of Avdakov as chairman and his replacement by Aleksandr Guchkov. As a gesture of conciliation toward Moscow, the TsVPK extended an invitation to Guchkov to serve as one of two new delegates to the Special Council.[73] On 23 June, Avdakov suffered what amounted to a vote of no confidence when his proposal to send representatives of the TsVPK to familiarize themselves with the activities of the provincial committees was rejected on the grounds that plenipotentiaries from the district committees to the central committee provided enough information.[74] Unable to halt the regionalization of the committee, Avdakov did not preside over another Presidium or plenary session.

Guchkov was in many ways the very antithesis of Avdakov. The latter was essentially an organization man, more at home in the board room or his office on Liteiny Prospekt (the Association's headquarters), than in the political arena. On the other hand, Guchkov, nominal chairman of the Octobrist Party and sometime president of the Third Duma, was a political activist *par excellence*.[75] Avdakov was the epitome of the petitioning industrialist, 'a latter day Kit Kitich' as Lenin called him; Guchkov has been characterized, not without admiration, as 'a liberal with spurs'.[76]

Born in Moscow in 1862 to a wealthy wool manufacturing family, Guchkov earned his spurs in such far-flung places as the Transvaal, Peking, Macedonia and Manchuria, in each case close to the battlefield if not actually on it. His military exploits, his 'will to danger', could hardly have been more alien to the experiences of his future colleagues on the TsVPK.[77] On the other hand, his subsequent career made him the natural, indeed, the ideal candidate to head any 'public' organization concerned with questions of military supply. As the chairman of the Third Duma's Commission on State Defence, Guchkov was a gadfly to the Ministries of War and Navy. His speeches occasioned by the annual debate of the GAU's proposed budget have a prophetic quality, which no doubt redounded to his credit during the war.[78]

'At the time of the last war', he told the Duma in 1908,

> We were compelled to get both cartridges and shells from abroad, and it should be clear to you that in the event of a European war, when such delivery will be impossible, we will find ourselves in a condition close to the catastrophic.[79]

A closed session of the Commission on State Defence in February 1912 heard him attack the supply system of the GAU in no uncertain terms.[80] In May, he concluded his report to the Duma on the GAU's estimates in the following manner:

> I must acknowledge that in five years of the closest familiarity with and study of this ministry's budget rarely did I come into contact with such a disorganized section as the Main Artillery Department....The simple transfer of an order to the appropriate technical office and thence to the corresponding factory takes on the average six months....We, in the course of the past five years, through constructive criticism and cooperation...tried somehow to inspire this section to perform its work more rapidly and rationally and if now...we turn to you with our cries of despair, if we tell you that without your indignant protest never will this evil be abolished, then believe me, this is truly an act of despair.[81]

With the outbreak of the world war, Guchkov, out of the political limelight since his defeat in the elections to the Fourth Duma, was in Warsaw with the Red Cross. He returned quite often to Moscow and Petrograd, supposedly to report on the supply and sanitary situation at the front, but according to Sukhomlinov, his real objective was to sow the seeds of dissent and plot for the minister's ouster.[82] In early June 1915, Guchkov turned up in Moscow prepared to give first hand accounts of supply shortages to any and all who would listen. His audience on 13 June consisted of Zemstvo Union officials, and his speech lasted two hours. Pares, who was present, described it as 'very remarkable'

and noted down as much as he could for the Foreign
Office. Guchkov enumerated several instances of
negligence in arms supply and held Sukhomlinov
responsible. More remarkable was his assertion
that while

> it was quite impossible to make the work of the
> existing artillery factories more intensive...it
> was possible to apply any mechanical means
> existing in Russia to the manufacture of ammu-
> nition as had been done in France since the
> beginning of the war. It was absolutely necess-
> ary to deprive of their monopoly factories which
> had so far enjoyed the exclusive favour of the
> Russian War Office, and regard [*sic*] any new
> output as a threat to this monopoly. If Russian
> industry applied itself thoroughly to the ques-
> tion, some months of work might even now bring
> the supply on to a satisfactory basis.[83]

These and other remarks by Guchkov endeared him as
much to his former political rivals in Moscow as
it alienated him from Petrograd banking and industrial circles. His subsequent rise in the TsVPK
signified among other things the ascendance of the
former group in the organization. His election at
the first congress in late July to the chairmanship of the TsVPK and that of A.I. Konovalov as
vice-chairman confirmed it.

3.4   THE FIRST ALL-RUSSIAN CONGRESS

The First All-Russian Congress of the VPKs
convened in the hall of the Society of Railway
Engineers in Petrograd on 25 July 1915. Bringing
together 231 representatives from 85 committees
plus scores of observers, it was essentially an
organizational congress, one that would pass
judgement on the largely *ad hoc* decisions taken by
the TsVPK in its first two months of operation.
Only the simultaneous debate occurring in the
State Duma over the government's proposal for a
Special Council of Defence (discussed in Chapter
4) managed to deflect the delegates' attention
away from internal questions.
     These concerned everything from the future
location of the TsVPK, to its social composition,

the distribution of responsibility among central, district and local committees, the procedures for designating district and local committees, and the finances of the organization. Barely concealed behind the discussion of these seemingly mundane questions was the struggle between those favouring the concentration of power in the hands of the Association of Industry and Trade and those wishing to recast the role of the industrial bourgeoisie.

If there were any doubts that the leaders of the Association intended to maintain their control over the TsVPK, these were dispelled at the third session of the congress on 27 July. Avdakov opened the debate with a reminder that the Association still exercised executive responsibility over the TsVPK and asserted that it would continue to do so at least until the organization had been officially recognized.[84] Zhukovskii, perhaps anticipating the latter development, urged that the TsVPK remain in Petrograd to be close to the levers of power. Listing 24 committees as candidates for district status, Zhukovskii sought to limit their powers to a determination of their organizational procedures and how best to prepare certain articles such as shovels, picks, axes and simple types of ammunition which could not be centrally controlled.[85] All orders were to come through the TsVPK as were advances granted by the state. The TsVPK was to deduct 1 per cent of the total cost of each order to cover organizational expenses. This, Zhukovskii assured the congress, would suffice.

Such a programme was unacceptable to many delegates for a variety of reasons. Some were aghast at the retention of power by the Association even for a limited period of time. A Professor Kossogovskii from Novocherkassk referred to the activities of the zemstvos and municipal organizations which also had no legal sanction and asserted that 'in time of war, it is not necessary to pay strict attention to legal questions'. The antipathy of many provincial delegates to the bureaucratic tendencies inherent in Zhukovskii's programme was typified by one delegate from Nikolaev who cried 'we know what needs to be done, and we from the periphery say to the centre we will do it'.[86]

Further objections were raised by M.I. Tereshchenko, the chairman of the Kiev VPK. Tereshchenko diplomatically thanked the Association for fostering the TsVPK and for extending its technical and financial assistance. But in his view the Association was an organization of peacetime. The VPKs were adapted for war and it was time for them to stand on their own. For this, if for no other reason, they should seek legalization immediately. Tereshchenko had little sympathy for Zhukovskii's justification for maintaining the TsVPK's headquarters in Petrograd. 'When we stand with the masses as one body, then we will be needed by the government and not it by us', he proclaimed. Though not a Muscovite, he preferred Moscow as the headquarters because it was 'closer to my heart in principle and spirit'.[87]

The delegates from the provinces viewed the Association as a front for Russia's raw material syndicates which were backed by Petrograd and foreign finance capital. In so far as many delegates represented nonsyndicalized domestically financed industries, or the consumers of industrial raw materials, they had more at stake in limiting the Association's powers than was involved in determining the nature of the TsVPK.[88] On the other hand, with the right leadership, the committee could serve as a counterweight to, if not a weapon against, the Association-syndicate axis. Tereshchenko went so far as to warn that if the syndicates constituted an obstacle to the goals for which the VPKs had been created, then it was the committees' responsibility to abolish them. But most delegates were not against the syndicates in principle. They merely wanted to prevent Prodamet, Produgol' and the others from taking advantage of wartime scarcities at the expense of their own industries. In the ensuing debate, it was Riabushinskii who reminded the congress that the source of Germany's industrial strength was its network of syndicates and cartels. Russia should not dismantle its syndicates, argued Riabushinskii, but place them in more 'responsible hands', those capable of 'changing the rules'.[89] Were these hands to be the VPKs? Riabushinskii did not elaborate.

On this and other points of contention between pro- and anti-Association forces an elaborate compromise was struck. As far as oil and coal were concerned, two commodities dearest to the Association, the resolutions adopted by the congress' section on raw materials were extremely favourable to the syndicates. The section's demands for the opening up of new land for oil exploitation and 'extreme measures' to supply coal mines with workers, blasting equipment and transport facilities were those that the oil trusts and coal syndicate had been making ever since the outbreak of the war.[90] Steel was treated differently. The congress rejected a motion for joint action with Prodamet on the preparation of statistics for orders, and included instead in its resolution a clause which stipulated that, 'the activity of the TsVPK must be directed towards the most rational distribution of orders to factories supplying defence enterprises with metal'.[91] This directive papered over, rather than resolved, differences among the committees' members about how to deal with steel and the syndicate which marketed it.

Zhukovskii's scheme for financing the committees' operations was adopted. On the other hand, the congress soundly defeated his attempt to limit the initiative of district and local committees. It granted them the right to purchase property, to enter into agreements with state as well as private institutions, to arrange for the fulfilment of orders as well as their delivery, and to answer in court any charges brought against them.[92]

In spite of their hearts, the majority of delegates voted in favour of Petrograd as the TsVPK's headquarters. However, this was not quite the TsVPK that Zhukovskii and other stalwarts of the Association had anticipated. Table 3.2 shows a comparison between their proposed TsVPK and that which finally emerged from the congress.

Thus, the Association had effectively lost control of the organization it had created. This revolt on the part of a sizeable proportion of the industrial bourgeoisie was a severe blow to the Association from which it did not recover. Its residual connections with, and tentative support for, the VPKs led to a breach with Petrograd's industrialists, which further weakened it. Most important in the immediate sense was the change at the top. Guchkov and his able assistant, Konovalov, now stood in place of the Association's council at the head of the TsVPK. Together with Riabushinskii, the chairman of the Moscow VPK, they constituted a triumvirate which was to leave an indelible mark on the organization. For better or for worse, they were to involve it in the high politics of defence.

TABLE 3.2   *Composition of Central War-Industries Committee*

	Number of representatives	
	proposed by Zhukovskii[93]	adopted by congress[94]
1. Elected by congress	chairman	chairman and vice-chairman
2. Association of Industry and Trade	10	10
3. District committee (each x total)	2 x 24 = 48	3 x 28 = 84
4. TsVPK sections	unspecified	unspecified
5. City and zemstvo unions	6	6
6. Moscow and Petrograd city governments	4	4
7. Military departments	unspecified	unspecified
8. Committee of Military-Technical Assistance	3	5
9. All-Russian Chamber of Agriculture	0	2
10. Workers	5	10
11. Individuals elected by TsVPK	5	8
	82	131

4 The High Politics of Defence, June–September 1915

The rapid industrialization of Russia in the late nineteenth and early twentieth centuries had willy-nilly laid the foundations for the formation of an indigenous industrial bourgeoisie. This process was not, however, accompanied by an aggrandizement of political power which its Western counterparts had obtained as part and parcel of their economic ascendance. During the first year of the Great War it became apparent that the state would have to draw on the economic strength which industrialists could command to furnish the necessary articles of war. But the Tsarist administration was loath to relinquish control over the deployment of the empire's economic resources of the country. The creation of the Special Council was a partial concession to private industry as represented by a 'commercial-industrial circle'. But the Petrograd financier-industrialists who comprised the circle were the least likely to demand political changes, having established a close partnership with the military departments.

The election of Guchkov and Konovalov at the VPKs' congress represented a step towards the politicization of the committees and a challenge to the existing state-industrial complex. Guchkov, Konovalov and Riabushinskii, the VPK triumvirate, were sufficiently alienated from the state bureaucracy to lead an assault on it. The attack took the form of a campaign to replace the existing administration of supplies with the Russian equivalent of a ministry of munitions, and was part of a larger campaign organized by the Progressive Bloc.

4.1 THE WAR-INDUSTRIES COMMITTEES AND THE SPECIAL
      COUNCILS

The government's policy with respect to arms
procurement changed considerably with the replacement of Sukhomlinov by Polivanov as Minister of
War and therefore chairman of the Special Council.
The large Petrograd metalworks factories were
irreplaceable as the nucleus of armaments
production, but their monopolization of it, aided
by Sukhomlinov, had caused the war effort and the
prestige of the government no little harm.
Polivanov sought to scale down their influence and
in this endeavour he found a willing ally in the
VPKs.
    At the first session over which Polivanov
presided, the number of representatives from the
commercial-industrial circle was six, five less
than during Sukhomlinov's brief tenure. Thereafter, it was no more than four, the number
prescribed in the Special Council's statutes.[1]
On the other hand, the contingent of those active
in the VPKs increased in number. Apart from the
three delegates approved before Polivanov's
accession, Fon-Ditmar attended as a representative
from the State Council; Guchkov and Litvinov-
Falinskii replaced two of the original VPK
delegates, but one of them, Avdakov, continued to
attend the Council's sessions also as a State
Council representative; finally, from 11 July,
Konovalov, appointed by the Moscow VPK, regularly
attended, bringing the total of VPK members to
six.[2]
    In addition to changes in personnel, the Special
Council after Sukhomlinov's departure began to
shift its procurement policy, largely at the
expense of enterprises represented by the commercial-industrial circle and in particular, the
Putilov Works. It was well known that Putilov had
a long backlog of unfulfilled orders. Indeed, as
early as the second session of the Special
Council, the commercial-industrial circle had
sought to have 'less urgent' (and not incidentally
less profitable) orders placed by the Naval and
Ways and Communications ministries before the war
either cancelled or transferred to other enterprises. Concerning the orders granted by the

Special Council under Sukhomlinov, representatives from Putilov announced that fulfilment on time would be difficult and that a large advance was necessary. At the same time, the Council received a complaint from the Khar'kov VPK that engineers dispatched by the Putilov Works to South Russia were trying to requisition machinery from a German-owned enterprise - closed by order of the state - and to induce its workers to sign on with Putilov.[3]

The Special Council was not about to cancel the orders to Putilov which Sukhomlinov had approved. But it did appoint a subcommittee to investigate the enterprise, placing the redoubtable Guchkov in charge. In his report tabled on 31 July, Guchkov revealed that the Putilov Works was heavily indebted to the state and private banks and that to resolve its financial difficulties it had systematically put off the fulfilment of old orders and frozen or actually reduced wages. Guchkov was also critical of the War and Naval ministries for assigning so many orders to Putilov without first assessing the productive capacity and financial condition of the enterprise.[4] The programmatic implications of the report were clear. The Special Council could no longer rely exclusively on Putilov or other Petrograd enterprises but had to encourage the development of war-industrial production by other means and in other places.

In fact, the Council had already taken steps in this direction. It began to furnish the public organizations with advances as a means of initiating munitions production among previously non-contracting enterprises. It approved 3 million rubles to the TsVPK, on 24 June for the purchase of machinery; in early July, it allocated 5 million rubles each to the Union of Towns, Zemstvo Union and Moscow VPK, plus 500,000 to the latter to stimulate production among small enterprises;[5] a further 3 million was granted to the TsVPK on 8 August; and it received 10 million more in the middle of the month.[6] Of a total of 43.3 million rubles assigned by the Special Council in advances, 31.7 million went to the public organizations, of which the VPKs' share was 21.5 million.[7]

The mechanical section of the TsVPK obtained its

first order on 13 July. The next day, four more were forthcoming from the GAU. By 19 August, the date of the last Special Council session, the TsVPK had received the following orders from the GAU and the Main Engineering Unit (hereafter GVTU):[8]

Article	Amount
Hand grenades	5,000,000
Time fuses for hand grenades	1-3 million/month
Bomb-launchers	2000
Shells	1,700,000
Mortars	500
Mortar shells	250,000
10 lb. high-explosive aeroplane bombs	5000
Parts to 4.8-inch howitzers	200
Parts to 3-inch mountain artillery	300

The committees, therefore, had benefited a great deal from the Special Council since Polivanov's accession. Their infiltration of the Council coincided with, and contributed towards, the breaking of Petrograd financier-industrialists' monopolistic hold over procurements, and led to an impressive outpouring of advances and orders to the committees. Such an outcome had indeed been predicted by *Utro Rossii*, Riabushinskii's newspaper and the Moscow VPK's unofficial organ until the appearance of its own *Izvestiia*:

> Undoubtedly the Special Council will depend on the work of the War-Industries Committees... because it would be impossible for it to enter into direct negotiations with the tens of thousands of large and small factories and workshops spread out across the country.[9]

But the possibility of the Special Council's
dependence on the VPKs alarmed the Council of
Ministers. Was not the Special Council a state
institution? Could the committees succeed in
making it their own organization? Short of abol-
ishing the Council, which, given the seriousness
of the munitions crisis and evidence of other
shortages, was out of the question, what could be
done? The submission of the Imperial rescript on
14 June to reopen the Duma not later than August
provided a solution. Since the Tsar had approved
the Special Council on the basis of Article 87, it
had to come before the legislative organs within
two months of their reopening or lapse. Rather
than submit the original statutes which the Tsar
had approved on 7 June, the ministers decided to
revise them thereby heading off the challenge
posed by the committees and various Duma fractions
which were in the process of formulating their own
proposals.

In early July, a report appeared in the press
that 'a special organization for distributing
orders under the Minister of War, consisting of
representatives of the Unions and a number of
industrialists' committees connected with a
specific product' was under consideration. This
organ would 'take over many of the VPKs' functions' and turn them into an 'auxiliary institution'.[10] This plan went through three drafts,
the final version being approved on 14 July.[11]

The exclusion of the committees' representatives
from the revised Special Council was a calculated
manoeuvre designed to deprive them of any leverage
vis-a-vis the military departments and private
industry. The ministers were, in effect, attempting to roll back the advances made by the
committees by enticing individual industrialists
away from them. At least in Moscow, this was
recognized and resented. 'Unfortunately', editori-
alized *Utro Rossii*,

there is no mention of representation of the
VPKs. Although leaders of the most important
Petrograd and Moscow committees will enter the
Special Council as representatives of the
exchange committees and the all-Russian industrial congresses, nevertheless the VPK appears
in practice as the main force organizing and

mobilizing industry to help the army, and to ignore it in this way cannot be considered correct.[12]

The plan which the ministers presented to the Duma, besides revising the membership of the Special Council, considerably expanded its area of competence. In addition to armaments production, it was to cover the supply of food to the military and civilian population, transport, and fuel. To this aspect of the plan there was no objection. *Utro Rossii* in the same editorial considered such an extension of control 'quite expedient'. The Progressist fraction's own formula for a Committee of State Defence covered virtually the same ground.[13] And the Kadet proposal for a Main Supply Administration, modelled on Lloyd George's Ministry of Munitions, was even less ambitious, excluding food supply.[14]

Where these projects genuinely differed was on the question of who should wield the power. The Progressists' proposed Committee, containing representatives of the Unions, workers, the legislative organs and the VPKs, would have decided matters on a collegial basis.[15] The Kadets' Main Supply Administration provided for the same representation minus those from the legislative organs, but a special manager (*glavnoupravliaiushchii*), empowered with ministerial status, would dictate decisions. The government's plan kept all executive authority in the hands of the Minister of War, and, as already noted, denied the VPKs even a consultative voice.

The Progressists abandoned their plan after it encountered the opposition of other parties, particularly the Kadets. P.N. Miliukov and A.I. Shingarev led this opposition, objecting to the plan's conflation of legislative and executive authority. The Kadet formula was submitted to the Council of Ministers and not unexpectedly was rejected as 'unacceptable in essence'.[16] It scarcely received better treatment at the two joint sessions of the Duma's Defence and Budgetary Commissions (23-4 July). The Commissions considered the creation of what amounted to a new ministry unrealistic in wartime. They criticized the government's plan too, but judging by their revised version presented to the Council of

Ministers on 28 July, had little sympathy for
industrialists' claim to be represented. The
Commissions eliminated all industrialists from the
proposed Special Council of Defence and instead
included an amendment which granted to the chairman the right to invite representatives from
the VPKs, scientific-technical societies, and
workers.[17]

The VPKs had an opportunity to comment on the
government's proposals and the Duma Commissions'
emendations at their first congress. P.P.
Riabushinskii and his business partner, A.I.
Kuznetsov, endeavoured to broaden the discussion
into an all out attack on the government. Both
emphasized that the existing regime was leading
Russia down the path of defeat. Kuznetsov
mentioned insidious German influence in Petrograd,
or 'Wilhelmgrad' as he mockingly referred to it.
'Gentlemen', he cried, 'it is impossible to live
any longer with this government. I firmly believe
in victory, but we must take matters into our own
hands.'[18] Riabushinskii accused the government
of weakness rather than treason.

we really do want a strong authority, but at the
same time we do not feel it. We are thus
perplexed. Who rules Russia at the present time?
If a serious answer could be given to this question it would have to be: we do not know.

Precisely because the government was weak, there
was no point in committing the VPKs to its
proposal for a Special Council of Defence.[19]
For, as he told a Moscow regional congress of VPKs
a few days earlier, 'the very system of government
must be fundamentally changed'.[20]

Most delegates, however much they applauded the
verbal gesticulations of Riabushinskii and Kuznetsov, focused their attention on the Duma Commissions' proposals and were appalled by the
exclusion of industrialists. The possibility of
consultative representation for the VPKs did not
diminish the delegates' anxiety. Iu.I. Poplavskii
equated the Duma's decision with the Moscow
pogrom, and promised that neither would intimidate
industrialists. Another delegate, either in jest
or spite, called for 50 representatives from industry to be included in the proposed council.[21]

The special council which the delegates envisioned, and which appears in the congress's resolution on the administration of supplies to the army, was really a 'statified' TsVPK. A special assistant Minister of War, a minister of munitions in everything but name, was to direct it. The congress's resolution describes him as 'a person invested with society's confidence'. The Odessa lawyer, M.S. Margulies, more accurately characterized him as 'a man from our midst'. Given 'the widest powers', the assistant minister could, in Margulies' somewhat embarrassing words, 'blow up the enemy from within'.[22] He would preside over all military departments which, to ensure maximum coordination of procurement policy, would merge into one administration.

This formula was the exact replica of that which A.I. Guchkov urged on the Moscow regional congress on 24 July.[23] The connection between Guchkov and the proposed assistant ministership actually was much closer. Among certain circles, he was considered a prime candidate for a ministerial position. P.B. Struve, for instance, regarded only Guchkov and Prince L'vov as capable of serving both state and society as ministers, and in a letter of 10 June told the Minister of Foreign Affairs as much.[24] At that time speculation centred on Guchkov's appointment as Minister of Trade and Industry, but by late July the press connected him with the proposed assistant Minister of War.[25] His election to the chairmanship of the TsVPK left little doubt about whom the committees' congress preferred.

Nevertheless, Guchkov had too many enemies within the state as well as 'society' to be accepted. The Kadets resented him ever since he had agreed to their exclusion from the Third Duma's Commission on State Defence. The record of the Kadet Party's central committee reveals the extent of this resentment. During its consideration of a plan for 'ministers of the public' in August, Guchkov's name arose because, as F.F. Kokoshkin remarked, 'it must be mentioned'. But Kokoshkin labelled him 'an element of division' and rejected him in favour of either Prince L'vov or Rodzianko.[26] Kokoshkin alluded to another problem in a letter, dated 26 July, to I.I. Petrunkevich, a fellow Kadet. Guchkov could not be a 'Russian Lloyd George' because 'his candidacy runs up against serious obstacles from above'. This was a reference to the court's well known hatred of Guchkov.[27] Polivanov, who was on close terms with Guchkov, felt compelled to avoid him in his selection of a new assistant (after the dismissal of Vernander) because of 'the Tsar's attitude toward Guchkov and toward the combination of Polivanov-Guchkov'.[28]

The Duma, in any case, rejected the idea of a special assistant minister. On the other hand, it regarded the multiplicity of tasks which the government had assigned to the Special Council of Defence as too complex for a single individual or even an entire ministry. The Duma Commissions decided to expand the number of councils from one to four, each to be placed under a different minister. This plan, approved by the State Council, gratified the legislators' sense of self-importance. Plenipotentiaries from the two chambers constituted 62 of 122 permanent members. At the same time, the government was undoubtedly relieved because executive authority was to remain in the hands of the ministers. Only the councils' chairmen could introduce questions for discussion, demand priority in the fulfilment of orders, designate sequestrated enterprises, appoint officials to investigate commercial records, resolve to close nonessential enterprises, and assign wage scales in those enterprises that were essential.[29]

The VPKs received some compensation from the Duma for the rejection of their formula in favour of a revised form of the government's. Between 28 July when the Commissions' report was sent to the Council of Ministers and 1 August when it was taken up by a full session of the Duma, the Commissions managed to add four TsVPK deputies to the list of members of the Special Council of Defence.[30] The TsVPK was permitted to send its chairman, Guchkov, a representative of science and technology, and two of the ten workers' representatives still to be elected. The Duma's full session, however, swept aside all such distinctions among TsVPK representatives by a vote of 141 to 138 on a motion of the Octobrist, Dmitriukov.[31] An attempt by M.M. Kovalevskii in the State Council to revive the curial basis for

TsVPK representation to ensure the presence of engineers and workers was beaten back by N.F. Fon-Ditmar who asserted that the committee in its wisdom would voluntarily include representatives from these groups.[32]
The VPKs fared worse in the Duma's deliberations on the other Special Councils. The Duma acknowledged fuel as a legitimate concern of the VPKs and accordingly granted. the central committee three places on the Special Council for Fuel. But it denied the committees any seats in the transport and food supply councils. The State Council duly approved these recommendations and they became law on 17 August.

The establishment of the four Special Councils meant that the VPKs could either accept a subsidiary role within the bureaucratic system or continue to struggle against it. Fon-Ditmar advocated the first approach. At the first session of TsVPK's Bureau following the congress, he expressed his annoyance with those who continued to criticize the government.

The weak side has already been revealed. There has already been criticism. Now, together with the departments we need to correct the shortcomings. Creative and not destructive work is what is required.[33]

Along with other representatives of the Association of Industry and Trade in the TsVPK, Fon-Ditmar opposed a political orientation if only because he feared it would cut off the hand that fed the VPKs.
What he failed to recognize, or did not want to admit, was that even if the TsVPK had refrained from criticism, its very existence would be seen by certain officials as a threat to theirs. The consideration of the committees' statutes, drawn up by the first congress, is illustrative of the mistrust which the VPKs inspired. General A.S. Lukomskii, the Assistant Minister of War, reported on 2 August in a memorandum to his superior that

The activities of the Central War-Industries Committee...have already demonstrated that it has an extremely liberal understanding of its rights and obligations. The committee addresses

its requests and inquiries not only to the
Special Council, but to separate departments of
the Ministry of War. I do not know whether the
committee has relations with other ministries
but it would seem in this respect desirable to
define in the statutes of the committees more
precisely their rights and obligations and not
to give them the character of a separate
ministry.[34]

Polivanov received another unflattering communication about the TsVPK a day later from General
M.A. Beliaev, the head of the General Staff administration. Beliaev cited six examples of what he
considered actions inappropriate for a non-
official organization. Among them were a
communique 'categorically' demanding that the
state designate the oil industry among those working for defence, a request to inform Russia's
purchasing agents in London and New York of
Russian industry's needs in machinery, a letter to
the General Staff advising it to evacuate Belostok
and to permit the town's Jewish population free
access to Russia's interior, and a telegram to the
head of the supply unit for the Southwest front
requesting that he send one engineer to Samara to
facilitate war production. The substance of these
petitions did not disturb Beliaev as much as their
'imperative tone' which implied that the Ministry
of War was to assist the committees and not the
other way around. He concluded his diatribe
against the committees with the identical characterization used by Lukomskii.

Veering sharply away from its assigned limits
...the TsVPK creates the impression of a new
central organ, approximating a ministry of
munitions endowed with extensive powers.[35]

Undeterred by his subordinates' reports, Polivanov
did his utmost to secure the approval of the VPKs'
statutes in full. It was thanks to his invitation
that Guchkov was able to address the Council of
Ministers on 4 August. According to A.N.
Iakhontov, who as assistant to the Chief of
Chancellery took notes of the ministers' sessions,
Guchkov demanded approval in full or a blanket
rejection of the statutes.[36] The ministers

eventually approved the statutes on 27 August, but
with several revisions. To Article 1 they added
the clause that the committees would only operate
'during the present war'. Article 2, which had
previously stipulated that the VPKs 'do not pursue
commercial aims', was amended to read that 'enterprises belonging to the committees will accept
orders at cost value'. Article 17 in its revised
form dictated that all VPK property would revert
to the state once the committees were
dissolved.[37]
  Apparently, the ministers did not object to the
VPKs as much as to the fact the Guchkov was their
chairman. According to the Minister of Ways and
Communications, S.V. Rukhlov, Guchkov was responsible for turning the committees into not merely a
new ministry but a 'second government' or 'revolutionary organ'. 'It is very sad', remarked A.A.
Khvostov, the new Minister of Internal Affairs,
'that this character has once more risen to the
surface and figures as head of the War-Industries
Committee' because 'it can be turned into a
dangerous weapon for the sake of political
games'.[38]

4.2   THE WAR-INDUSTRIES COMMITTEES' TRIUMVIRATE AND THE PROGRESSIVE BLOC

The Progressive Bloc, a coalition of all the Duma
fractions except the extreme right and the extreme
left, arose in August 1915 during the waning days
of the Duma's third session. It existed in impotence until late 1916. The Bloc has received much
attention from historians since the publication in
the early 1930s of material relating to its
origins and various meetings held pursuant of its
aims.[39] Analysed in terms of pre-war political
alignments, it represented, according to one
interpretation, an end to the 'third of June
system' inspired by Stolypin, and to another, the
perpetuation of that system under different
circumstances.[40] Viewed through the prism of the
events of 1917, it has been described by one of
its leading architects, Miliukov, as an attempt to
give 'the last chance to the Tsar for saving the
dynasty in Russia', and by a Soviet historian as
'the antithesis of revolution', which is really

arriving at the same point from a different angle.[41]

The main characteristic of the Progressive Bloc during its unhappy existence was ineffectualness. The right-wing Kadets, unwilling to admit to themselves the basic conservatism of their liberalism, rationalized their timidity with mixed, long-winded and borrowed metaphors. Russia was an 'inflamed wound' which could be set alight by 'one carelessly thrown match' (Miliukov). Russia was that automobile driven by a reckless chauffeur and full of those who saw the need to change drivers but were afraid to attempt it lest they throw the vehicle into the abyss (V.A. Maklakov). Russia was a rider who could not afford to change horses in midstream (Miliukov).[42] The Unions' leaders, Chelnokov, Prince L'vov, N.M. Kishkin and N.V. Nekrasov, were content to concentrate on the practical work of their organizations which would 'serve as a lesson of huge significance'.[43] 'We will not enter on the path of a parliamentary struggle in war-time', declared Prince L'vov in August 1915. This would only 'give occasion for the government to cut down on our work to which we are more responsible than to the political movement', he added at a session of the Bloc on 25 October.[44]

Strictly speaking, the War-Industries Committees did not belong to the Bloc. Rather, the two movements were symbiotic. The committees required sympathetic ministers as much as the Bloc needed ministerial candidates without whom its slogan, 'a government of public confidence', would have appeared even more nebulous than it really was. The list of proposed ministers which was published on 13 August 1915 satisfied both.[45] Guchkov now appeared as a prospective Minister of Internal Affairs, and Konovalov was the inevitable candidate for (and indeed eventually became) Minister of Trade and Industry.

On one level, the VPKs' triumvirate, quite apart from the committees, hosted meetings attended by Moscow notables from the commercial-industrial as well as professional and local political worlds. Riabushinskii presided over such a gathering on 13 August which prepared a resolution adopted in the Moscow city duma five days later. The resolution, the first of many approved by city dumas and

exchange committees throughout the country,
demanded 'the calling of a Government strong in
the confidence of the public and unanimous, at
whose head must stand someone in whom the country
believes'.[46] On 16 August, Konovalov's apartment
was the venue. The topic of discussion was a
network of coalition committees consisting of the
three main public organizations. The coalition was
supposed to work for a 'cabinet of national
defence'.[47]

On a second level, the triumvirs used the VPKs
as forums for political discussion and rallying
points for the Progressive Bloc's campaign. Such
was the significance of an extraordinary meeting
(25 August) called by Riabushinskii and attended
by over 100 representatives of 34 VPKs from the
central industrial region. The resolution which
the meeting approved spoke of 'the inevitability
of the reorganization of the government's leadership and the necessity of immediately coopting new
people who possess society's trust'.[48]

The committees' leaders justified their participation in the overtly political campaign and the
political resolutions passed by the VPKs on the
grounds that the government as then constituted
provided no opportunity for the organization to
fulfil its promise to mobilize Russian industry.
Riabushinskii, for example, declared at the extraordinary meeting

> We have a right to demand that we be given every
> possibility to work....There is a danger that we
> will be held responsible where we are not.
> Therefore, for the sake of such responsibility
> we must turn our attention to the very system of
> governmental authority.[49]

Guchkov at a TsVPK Bureau session the day before
complained that 'as a result of the unhelpfulness
and disorganization of the authorities, our
activity is not accompanied by such productiveness
as would be desirable'.[50] But Riabushinskii and
Guchkov might have saved their explanations and
excuses, for already events were making them
irrelevant. The Tsar, the object of their entreat-
ies, had had enough. His so-called liberal phase
had come to an end. His last conciliatory gesture
was to preside over the ceremonial opening of the

Special Councils at the Winter Palace on 22
August. He read a declaration composed by his most
liberal ministers (Polivanov and Krivoshein),
listened to the oblique criticisms of his government made by Rodzianko, accepted a far more
forthright memorandum submitted by Shingarev on
behalf of the Duma's Commission on Defence, and
generally, in the words of the British ambassador,
'made an excellent impression on the committees'.[51] On 23 August, convinced of the
impotence of liberal protest, he left for Headquarters to replace the unreliable Nikolai
Nikolaevich as Supreme Commander of the Army.[52]
    All that remained was the dismissal of the
bothersome Duma. Konovalov and Riabushinskii
pessimistically predicted that this would come
with the conclusion of the Russian Finance
Minister's negotiations in London. In such an
eventuality, argued Miliukov, the Duma should make
a special appeal to the Tsar. Konovalov and his
party colleague Efremov disagreed. They wanted the
Duma to continue its sessions and appeal to the
masses for support.[53] This was, in fact, the
action which the Duma took, not on 3 September
1915 when with a whimper it complied with the
Tsar's rescript, but on 27 February 1917.
    The prorogation of the Duma provoked a new round
of meetings. Among the Moscow business community
and intelligentsia, ten were held on 5 September
alone, according to police reports. Here and there
could be heard calls for militant action. The
police reported that Riabushinskii and Poplavskii
were pushing for a strike by the public organizations. Efremov urged the same at a meeting held a
few hours after the prorogation.[54]
    But the Tsar's rescript also intensified the
wave of strikes at Russia's leading armaments
factories. The Putilov Works was struck on 2
September followed by other major plants in Petrograd. Factory inspectors reported that 34
factories, containing 64,000 workers, were
affected.[55] In such circumstances, the Progressive Bloc's members did not mince words. Maklakov
declared that he wanted no part of the overthrow
of the government if strikes were involved.
Chelnokov denounced a protest strike of tram
drivers in Moscow.[56] During the fourth
congresses of the Union of Towns and Zemstvo Union

(7-9 September) he and Prince L'vov were instrumental in turning away a delegation of 70 workers which sought to gain entrance. To the assembled delegates they urged 'patience and fortitude'.[57] The Unions officially associated themselves with the campaign for a government of public confidence but rejected more radical resolutions.

The VPKs' leaders also succumbed to the panic. Guchkov explained at the Union of Towns' congress that

> not for revolution do we call on the authorities to agree with the demands of society, but precisely for the strengthening of authority and to defend the country against revolution and anarchy.[58]

Riabushinskii, addressing the Moscow Exchange Committee, recommended a variation of the Miliukov plan.

> We must peacefully insist on the necessity of reviving the work of the Duma...We must elect our representatives and with those of the Unions and the War-Industries Committees go to the Tsar to tell him what is necessary.[59]

Such a delegation may have satisfied the monarchist sentiments of Prince L'vov, Chelnokov and others commissioned by the Unions, but not the monarch who refused to give it a hearing. The Progressive Bloc which had prostrated itself so low that it could go no further, temporarily dissolved itself, and with it, the hopes for a government of public confidence.

The defeat of the Bloc's campaign clarified the situation of the VPKs a good deal. Despite the exaggerated fears of the Minister of Ways and Communications, they had not become a second government, and had not managed to alter the policies of the one in existence. Despite the ambitions of the committees' triumvirs, they did not appropriate the administration of the war economy. They had not become a ministry of munitions, or even part of one, but, ironically, a tool of the Special Councils. In this capacity, the VPKs became intricately involved in the production of military supplies.

5 The War-Industries Committees and the Production of Military Supplies

The importance of the state to the Russian economy was the axiom of its industrial development. Since the period of extensive railway construction in the 1890s the state's requirements had constituted the chief demand for many of Russia's heavy and extractive industries. This demand manifested itself in the form of state orders.

Almost from its inception, the Association of Industry and Trade addressed itself to problems arising from the dependence of private industry on state orders. The main difficulty, as the Association saw it, was that the dependence was not mutual. Private industry had practically no other market to which it could turn. The plants generated in the hothouse atmosphere of state protection and state demand had trouble surviving outside the hothouse. As one official of the Association admitted in reference to the metallurgical and machine construction industries, 'we have a situation in our country where without the state it would be impossible to live'.[1] The state, on the other hand, could, when expedient, develop its own enterprises or distribute orders to foreign concerns.

Against both these alternatives the Association waged a continual battle. State entrepreneurship was decried as tantamount to 'state socialism'. To send orders abroad was considered bad for the balance of payments, dangerous to national defence and, on both these counts, something unpatriotic.[2]

To improve its position $vis\text{-}à\text{-}vis$ state and foreign competitors, organized private industry in Russia devoted itself to the reform of procedures for the distribution of state orders. A number of issues were involved:

> Under procedures based on a statute enacted in
> 1830, not only were government orders and
> contracts allocated at auctions in which state
> enterprises competed with a selected number of
> private entrepreneurs, but a lack of standardiz-
> ation in the conditions imposed on private
> contractors provided the basis, it was asserted,
> both for official extortion and for other
> exploitative exactions on private industry.[3]

In the course of one year, 1909, the Association
hosted at least four meetings concerned with the
state's policy of distributing orders for railway
and naval equipment to its own or foreign enterprises.[4] In addition, the fourth congress of the
Association drew up 'Statutes on State Supplies
and Works'. The main proposal embodied in the
statutes' 177 articles was the creation of an
interdepartmental economic committee with representatives of industry which would determine the
annual level of state orders and approve all
orders intended for foreign fulfilment.[5] This
recommendation was taken up by the Kharitonov
Commission (see p. 43) and discussed several
times, though never adopted, by the Council of
Ministers.

The war, of course, led to an increase in the
amount of state orders but only intensified
problems connected with their distribution. The
state above all required the fulfilment of orders
for military supplies in as short a period as
possible. The problem was that it had no policy to
encourage this. As the circle of state contractors
expanded, taking in textile and leather manufacturers previously oriented towards the market, the
Council of Ministers passed a number of resolutions. However, these were either inappropriate
to the crisis or impossible to enforce. An edict
of 4 September 1914, for example, gave priority to
orders from the Ministries of War and Navy but
allowed for exceptions in the case of fuel and
materials for military contractors, thereby
leaving the system practically unchanged. Another
provision entitled the state to requisition
supplies and sequestrate the property of enterprises refusing to accept orders for military

supplies, but as the military departments had inflated their prices already, the threat was meaningless.[6] Punitive measures, in any case, could not compensate for the lack of a comprehensive plan.

5.1 THE DISTRIBUTION OF STATE ORDERS

Temperamental and political differences among the VPKs' leaders did not entirely obscure their unanimity with respect to the improvement of procedures for the distribution of state orders. Before the war, the Association of Industry and Trade had frequently petitioned the government, and criticized it when it failed to respond. The VPKs, spurred on by the supply crisis and the challenge which Riabushinskii had flung at the government, looked to resolve the question in their own way.

The solution which the TsVPK initially proposed was quite simple. At one of its first sessions, it decided 'to concentrate the distribution of all orders in Russia as well as abroad in the TsVPK'.[7] This was intended as the Russian equivalent of an agreement concluded between the German government and the *Kriegsausschuss der Deutschen Industrie* in January 1915. As reported by *Promyshlennost' i torgovlia* (Industry and Trade) in its 1 June issue, the government gave its pledge not to distribute orders to individual enterprises.[8] Rather, they would go to 'distributional centres' such as *Kriegsrohstoff-abteilung* (Raw Materials Section), *Kriegsmetall*, *Kriegschemikalien*, and others 'staffed by men closely connected with the industries using the... materials which were to be controlled'.[9]

In Russia, no such agreement emerged. In fact, the spectre of the monopolization of orders by the VPKs inspired the Council of Ministers to seek their exclusion from the proposed Special Council of Defence. The committees' counterproposal to combine all military departments and place them under a special assistant Minister of War can be seen as a departure from the central committee's 'decision' of June but not from the aim embodied in it. The setback which the campaign for a government of public confidence suffered in

September, however, disabused the TsVPK of this
illusion. It thereafter renounced any intention of
monopolizing orders.[10] The idea was kept alive
by several provincial committees, and shortly
after the February Revolution - when hopes for
fruitful relations between the committees and the
military departments were highest - the Moscow
committee included it in a report on how to
improve the financial situation of the organization.[11] But these were examples of wishful
thinking rather than serious proposals.

The VPKs accepted whatever orders were forthcoming from the military departments and garrisons
and provided technical assistance, equipment and
raw materials to those enterprises enlisted to
fulfil them. How many orders did the committees
distribute? What kind of orders were distributed,
and to which enterprises? Concerning these questions and others related to the VPKs' role in the
war economy, there has been much confusion.

Part of the confusion stems from the incomplete-
ness of available data. The TsVPK issued monthly
accounts of its budget but was less thorough in
revealing the sum of orders which passed through
it. The Moscow committee was more conscientious in
the latter respect but Moscow was only part of the
system. Rarely did either the central or Moscow
committees list the amount of orders fulfilled.
Reports from other committees were sporadic and
imprecise. The VPKs' own figures can, to some
extent, be compared with and supplemented by those
of the Special Council of Defence whose Supervisory Commission undertook a number of
investigations of the committees. Finally, a
comparative study of orders processed by the TsVPK
and the Zemgor published by the latter
organization, and a publication of the (Soviet)
Central Statistical Administration are of some
value, though they must be used with special
caution.

The amount of orders in rubles distributed by
the TsVPK is discernible for the period from
September 1915 to August 1916 (Table 5.1). Data is
practically nonexistent for the last months of the
Tsarist government as well as the post-February
period. Subsequent estimates of the total amount
of orders distributed vary from under 200 million
to at least 800 million.[12]

TABLE 5.1 *The distribution and fulfilment of orders by the Central War-Industries Committee**

Date	Total distributed**	Increment**	% Change	Total fulfilled**	Increment**	% Change	Rate of fulfilment
1 Sept. 1915	36.69	–	–				
1 Oct. 1915	118.40	81.71	222.70				
1 Nov. 1915	169.02	50.62	42.75				
1 Dec. 1915	199.89	30.87	18.26				
1 Jan. 1916	211.02	11.13	5.56	1.81	–	–	
1 Feb. 1916	232.44	21.42	10.14	2.30	0.89	47.51	00.85
1 Mar. 1916	229.39	-3.05***	1.33	7.23	4.93	214.30	00.98
1 Apr. 1916	248.03	18.64	8.12	14.02	6.79	93.86	3.98
1 May 1916	261.35	13.32	5.37	18.29	4.27	30.50	5.64
1 June 1916	280.84	19.49	7.43	23.07	4.78	26.23	7.00
1 July 1916	306.46	25.62	9.12	32.70	9.63	41.85	8.21
1 Aug. 1916	320.40	13.94	4.55				10.01
1 Jan. 1917	390			132			33.74

*   Includes only orders of the GAU, GIU, and GVTU.

**   Figures in millions of rubles.

***   Indicates that amount cancelled was greater than amount distributed.

Sources: For total distribution see *Doklad schetnogo otdela na 26/II, 4/VII, 26/IX 1916*, three separate publications (Petrograd, 1916) pp. 13, 21, 21 respectively; for total fulfilled from 1 Jan. 1916 to 1 July 1916, 'Sdano vedomstvym', TsGVIA, f. 369, op. 1, d. 206, l. 273; for 1917 figures see A.L. Sidorov, *Ekonomicheskoe polozhenie Rossii v gody pervoi mirovoi voiny* (Moscow, 1973) p. 203, citing TsGVIA, f. 13251, op. 1, d. 35, l. 74.

According to the central committee, it had distributed orders valued at 395 million rubles by February 1917, 425 million rubles by October of the same year, and 600 million rubles by January 1918.[13]

In comparative terms, the Special Council of Defence's Supervisory Commission reported the following:[14]

TABLE 5.2   *Distribution of wartime orders, July 1914–May 1916*

Department	Total amount of orders (rubles)		Amount to TsVPK (rubles)
	July 1914– July 1915	*July 1915– May 1916*	
GAU	1,284,336,286	2,316,865,590	128,310,317
GIU	2,101,827,511	1,556,914,970	100,772,970
GVTU	280,986,360	349,880,591	30,295,127
Medical	28,628,842	20,805,843	1,975,143
All depts	3,695,778,999	4,244,566,994	261,352,879
Total	7,940,345,993		

The TsVPK obtained 3.29 per cent of all wartime orders distributed to 1 May 1916, but 6.2 per cent of those distributed since it commenced operations. It should be noted that the figures for orders distributed by the departments do not distinguish between domestic orders and those granted to foreign enterprises. Since the TsVPK was almost entirely oriented towards domestic production, the comparison is therefore biased against it. If we compare the 180–5 million rubles in orders which the central committee obtained in 1916 with the sum distributed by the military departments to domestic enterprises, the percentage climbs to 7.6.[15]

The TsVPK's orders were supposed to be distributed among district and local committees based on the latter's reports of prevailing conditions

and prospects for production. The provincial
committees would in turn approach enterprises
located within their region and negotiate
contracts. In actuality, every link in this chain
was weak. Committees often encountered difficulties in obtaining from enterprises precise
information on their productive capabilities. It
was all too easy for an entrepreneur to over-
estimate his plant's capacity to obtain an order,
and then having received it, exaggerate the difficulties in fulfilling it. In many instances and
areas of the country, the difficulties needed no
exaggeration. Expectations of supply deliveries
were frequently unjustified and had to undergo
repeated revision. Provincial committees did not
always report what they knew. Consequently, the
central committee was inclined to circumvent
district and local committees.

During the formative stage of the organization,
the central committee could do nothing but negotiate directly with industrial enterprises. But
what had begun as a necessity soon became a
convenience. Already on 31 August 1915, a provincial plenipotentiary demanded the cessation of
this practice.[16] Nevertheless, by December,
provincial committees had placed only 89.4 million
rubles of the TsVPK's orders while those distributed directly to private enterprises amounted to
110.4 million. Later that month, in discussions
with the central committee's Bureau, representatives of the Kiev and Tula committees voiced
identical complaints about the relative neglect of
the organizational network. They claimed that they
were being deprived of a source of income, the 1
per cent commission charge. More seriously, they
wondered about the necessity for their committees
to exist at all.[17] The second congress in
February 1916 resolved that 'only in cases of
special urgency or for orders for special items
the production of which is concentrated in a small
number of factories' could the central committee
bypass the provincial committees. But this, like
most of the congress' resolutions, was not carried
out. By August 1916, the proportion was nearly
unchanged, with orders distributed directly
amounting to 163.8 million rubles and those
through the provincial committees to 156.5
million.[18]

The central committee further insulted the provincial committees by placing the largest orders itself. Approximately four out of five contracts concluded by the VPKs went through provincial committees, but, as already implied, the fifth was larger than the other four combined.[19] According to the mechanical section of the central committee, orders worth 62.8 million rubles had been placed directly with 175 enterprises by February 1916. This was an average of 369,000 rubles per enterprise. Through provincial committees were distributed orders valued at 66.26 million rubles to 580 enterprises, or 114,000 rubles per enterprise.[20]

While some committees complained of unutilized capacity in their areas, others, not so much in defiance of the central committee as independent of it, obtained orders from elsewhere. Both the Kherson and Elizavetgrad committees admitted that they had accepted orders from the military departments, which was a violation of the organization's statutes.[21] The Moscow VPK which did have the right to accept such orders made extensive use of it. As of February 1916, the committee was granted 82.19 million rubles in orders of which only 10.39 originated with the TsVPK.[22]

Although the bulk of orders – 55.06 million rubles – was obtained from the GIU, this represented only a small fraction of Moscow's textile production for the army. Far more significant was the assistance of the committee's textile sections in allocating orders. Continuing the work begun by the exchange committee and other commercial-industrial organizations in 1914, the Moscow VPK's cotton, linens and wool sections distributed orders to textile factories, arranged for deliveries of raw materials and fuel, organized the delivery of the finished product, in short, did everything but sign the contract. This, the enterprises concluded directly with the GIU. Orders thus handled by March 1917 included 2319 million arshins of fabric and smaller quantities of ribbon, bandage and braiding. The general value was 755 million rubles.[23] Of this, only 360.7 million arshins of fabric and two million of ribbon had been distributed before the establishment of the textile sections of the Moscow VPK. The sections claimed to have attracted 347 firms

to defence work. Forty-two each fulfilled more
than 10 million rubles in orders.[24]
Another source of orders were local military
offices and units. The Viatka district committee
reported in February 1916 that it had received
806,600 rubles in orders from the TsVPK but
additionally concluded contracts with the provincial office of the GIU for 315,000 rubles. The
Novonikolaevsk local committee of Tomsk province
endeavoured to supply local garrisons with meat
and salt, and these articles represented 800,000
of a total of 1,446,375 rubles in orders.[25]
On a still larger scale, many committees
accepted orders from the Zemstvo and Town Unions.
By April 1916, 24 VPKs were on the Zemgor's
accounts.[26] The Vankov Organization also distributed orders to various committees for 3-inch
shells. Among the major recipients were the Kiev
and Ivanovo-Voznesensk committees (525,000 shells
each); the Serpukhov local VPK which distributed
them via N.N. Konshin's giant textile plant to
other converted enterprises in the vicinity; and
the Odessa district committee (410,000).[27]
Finally, committees close to the front-line
units undertook to provide them with stores which
might have been more difficult or taken longer
to obtain through normal military procurement
agencies. Orders from units at the front to the
public organizations reached a level in mid-1916
which, so the Ministry of War contended, disrupted
the logistical calculations of the departments.
The Ministry issued requests to cease this circumvention of normal channels but without much
success.[28] Following the delivery of a report to
the TsVPK in May 1916 which indicated that the
Unions had received much favourable publicity as a
result of their activities at the front, those of
the VPKs were stepped up.[29] In July, Baron
Maidel was sent to the Northwest front in search
of orders. In September, he toured the Caucasian
front and returned with a list of items to be
produced directly for the units stationed there.
Such activity increased in the months before the
February Revolution and continued thereafter.[30]
All these examples testify to the resourceful-
ness of the VPKs in the face of a disappointing
supply of orders from the military departments and
the central committee. They also support the

notion that it is virtually impossible to determine with any accuracy the total value of orders processed by the organization. The question of orders, however, can be pursued according to an additional criterion, that of the type of order distributed by the committees.

By March 1916, the VPKs had negotiated 798 contracts for orders obtained through the central committee. Of these, 447 (56 per cent) were for the GAU, 202 (25.3 per cent) for the GIU, 80 (10 per cent) for the GVTU, 4 (0.5 per cent) for the Shipbuilding Unit of the Naval Ministry, and 65 (8.2 per cent) were for articles ordered by the TsVPK itself.[31] The regional distribution of these contracts reveals considerable variation in their average value, but no clear pattern emerges in this respect (Table 5.3). The preponderance of Petrograd in terms of number of contracts suggests that the central committee valued the convenience of dealing with local enterprises above regional development.

The 798 contracts covered 116 articles. The mechanical section of the TsVPK was responsible for 441 contracts pertaining to 55 items.[32] The most frequently ordered articles, their total cost and the number of enterprises producing them as of February 1916 is presented in Table 5.4.

As impressive as the figures in the table may appear, they are of little significance out of context. It is the *relative* importance of the VPKs in satisfying the army's requirements with which we are most concerned. Unfortunately, a direct comparison with the total amount of articles ordered by the GAU by the same date is not possible with the available data. We can, however, compare what the army expended in the course of the war (1914–18) with what was delivered by the VPKs by January 1917. According to the Central Statistical Administration, the army used approximately 48 million grenades, presumably of all varieties. The same source indicates that the TsVPK obtained orders for 16.56 million grenades, of which 14.33 million were delivered.[33] This would account for 29.8 per cent of the total expended. That not all grenades were expended by 1918 is more than compensated by the fact that the period of expenditure covers one year during which the VPKs did not exist (July 1914–July 1915) and another for which data is not provided (January 1917-January 1918). Moreover, the archival source used in Table 5.4 lists as ordered 5.23 million hand grenades of the 1914 type (tin-plated) in excess of the Statistical Administration's table. Finally, the latter seriously underestimates the total value of orders to the TsVPK as well as the amount delivered. Another source, General Manikovskii, gives 30 million as the quantity of grenades delivered to the army from 1 January 1915 to 1 January 1917.[34] Taking the figure of 14.33 million as the TsVPK's share, this would account for an impressive 47.7 per cent.

TABLE 5.3 *Regional distribution of contracts for articles ordered through the Central War-Industries Committee, 1 March 1916*

Area	Number of contracts	% of total	Value of contracts*	% of total value	Value per contract
Caucasus	28	3.5	15.28	7.7	546 000
Ekaterinoslav	42	5.3	8.23	4.2	195 000
Khar'kov	49	6.1	11.15	5.7	227 000
Kiev	22	2.8	9.83	5.0	447 000
Moscow	92	11.5	32.15	16.3	349 000
Nizhnii Novgorod	49	6.1	6.01	3.1	122 000
Odessa	96	12.0	17.07	8.7	177 000
Petrograd	263	33.0	60.87	30.7	231 000
Revel	15	1.9	2.95	1.5	196 000
Rostov-on-Don	38	4.8	8.69	4.4	279 000
Siberia	37	4.6	10.63	5.4	287 000
Urals	67	4.4	14.30	7.3	213 000
Total	798	100	197.24	100	

* In millions of rubles.

Source: Statisticheskaia razrabotka, pp. 9, 12.

TABLE 5.4   *Main articles ordered by the
mechanical section of the Central
War-Industries Committee, 1 February 1916*

Article	Quantity	Total cost	Number of enterprises
9-cm bomb-launchers	7,880	1,550,000	58
9-cm shells	3,261,000	7,107,000	146
Hand grenades (1914 type)	18,732,905	39,751,760	95
Hand grenades (German type)	3,020,000	6,268,500	20*
58-mm mortars	2,360	665,000	5
Shells for 58-mm mortars	692,000	15,034,270	23
3-inch shells	500,000	950,000	4

* As of 9 Oct. 1915 according to 'Vedomost' priniatykh i raspredelennykh zakazov dlia ruchnykh granat, germanskogo obraztsa', TsGVIA, f. 369, op. 4, d. 117, 1. 27.

Sources: TsGVIA, f. 369, op. 4, d. 20, 1. 64obv.; TsGVIA, f. 369, op. 21, d. 49, 1. 50.

Similar comparisons for 58-mm mortar shells and 9-cm shells suggest that the TsVPK accounted for 10 and 14.7 per cent of the total expended according to the Statistical Administration, and 39 and an incredible 70 per cent according to Manikovskii. Still another source demonstrates that of the 4534 58-mm mortars produced in Russia in 1915-17, the VPKs accounted for 1776 or 39 per cent.[35] These figures, though by no means definitive, imply a greater reliance on the VPK network for particular items than an assessment of the value of undifferentiated orders would show.

5.2   THE CONTRACTORS OF THE WAR-INDUSTRIES COMMITTEES

Theoretically, the VPKs had authority to determine

with which enterprises they could sign contracts as well as the conditions stipulated therein. The exercise of these quasi-official powers was, however, severely constrained by the military departments' technical specifications and schedule or terminal date of fulfilment, as well as by the preference of some prospective contractors to negotiate directly with military officials.

At first, it was generally believed that the committees would concentrate on large-scale enterprises.[36] The TsVPK conceived of a division of labour between the committees and the two Unions whereby the latter would mobilize small-scale and cottage industry while the former would organize the production of cannons, shells, rifles and 'articles that can be fulfilled by factories the owners of which will enter the VPKs'.[37] Such an arrangement was obtained in Kiev with the creation of a joint committee for supply of the army. The committee's chairman, the venerable A.A. Bobrinskii, informed the Special Council that

> articles whose production is possible in large-scale factories should be referred primarily to the War-Industries Committee while medium and small-scale artisan and cottage industry should be under the Unions.[38]

These expectations and instructions may well have satisified the *amour propre* of the VPKs' officials, but they hardly succeeded to entice owners of large factories or military officials, neither of whom had the slightest intention of relying on the committees to process orders. As S.O. Zagorskii, writing in the 1920s, noted,

> The big firms had old-established connections with the different government offices, were better supplied with materials and machinery than the small firms, and in general were better placed on the market. They were in no need of help from the Committees. They had no wish to be taken under control, or to be in any way restricted in their business. The military bureaucrats and the officers of the Special Council received active support from these firms in their policy of retarding and thwarting the War-Industries Committees.[39]

Orders for rifles, machine-guns, heavy artillery pieces and other ordnance articles requiring large capital investments were not entrusted to the committees. The central committee's sole venture into artillery production by the major war-industrial firms proved abortive. In August 1915, it approached a number of state and privately owned concerns with the object of sponsoring the collective production of 250 3-inch field guns. A meeting of directors thereafter drew up the division of responsibility: the state's Obukhov and Izhorsk factories would cast and forge the required pieces; the Franco-Russian and 'Phoenix' machine construction plants, the Russian-Baltic Shipbuilding and Engineering Works, and the soon to be opened 'Reikhel' Works would provide gun-carriages; and the Putilov Works would assemble and test the guns. All work was to proceed so as not to interfere with other orders. On this condition, the plan received the approval of the Special Council of Defence's Preparatory Commission, and on 21 October the Special Council of Defence granted an advance of 3 million rubles. However, before work progressed beyond the initial stage, several factories, claiming prior commitments, pulled out. A subsequent memorandum from the TsVPK to General Lukomskii suggested that the GAU redistribute the order and promised to return the advance.[40]

Much of the literature dealing with the VPKs casually refers to their mobilization of middle-range and small-scale enterprises which supposedly relegates their participation in war-industrial production to insignificance.[41] Such a characterization must be qualified, if not rejected, on three counts. First, what constituted a small as opposed to a medium or large-scale enterprise depended on a number of criteria - size of labour force, amount of fixed capital, branch of industry and even geographical location. Secondly, while most of the orders placed by provincial committees went to enterprises which by any of these criteria were not large, this was less true of those contracting with the Central and Moscow VPKs. Finally, the task of mobilizing enterprises, that is, converting them from production for the market or of 'non-essential' goods to war production, had dimensions which transcended the purely quantitative.

In the case of orders distributed by the TsVPK, a disproportionate share went to a small number of firms. Of 142 enterprises to which the TsVPK distributed orders for 13 million hand grenades in the autumn of 1915, 13 were responsible for the production of 5.45 million.[42] Five enterprises in the Baltic area obtained orders for 34 per cent of the committee's consignment for bomb-launchers, though they comprised only 10 per cent of the total number of enterprises involved.[43]

Among these enterprises were some of the giants of Russian heavy industry. The Nevskii Shipbuilding and Engineering Works with fixed assets worth 7 million rubles, produced bomb-launchers, 3-inch shrapnel and mortars for the TsVPK. The Mal'tsov Company, which accepted orders for mortars and shells, operated 11 factories throughout Russia with fixed assets totalling 12 million rubles. But whereas the orders to these and other large-scale enterprises were significant for the TsVPK, they appeared diminutive in the companies' books. The Putilov Works alone accepted orders from the state valued at 82.4 million rubles in 1915 or almost half of the total distributed by the committee as of the end of the year.[44] On the other hand, for enterprises such as 'Grammofon' of Petrograd, 'Zhest'' of Rostov-on-Don, 'Sirius' of Riga, or Kramer and Son of Mitau, the orders obtained from the TsVPK and the committee's assistance in securing machinery were of considerable importance. These enterprises, each employing several hundred workers and deriving their fixed capital primarily from domestic sources, were all new to war production and lacked 'old-established connections'.[45]

The situation was slightly different in Moscow where the region's leading entrepreneurs played an active role in the VPK. The committee's standing among textile manufacturers was quite high, and thus the choice of its textile sections as the expediters of the GIU's orders was a logical one. The committee also capitalized on its connections with metalworks enterprises. An outstanding example of this were the efforts of P.P. Riabushinskii to use the committee's shell section as a conduit of orders for the Moscow War-Industry Company. As he wrote to his brother Stefan, a member of the company's board of directors, this

> has for all of us special significance since it
> stimulates the activity of the VPK without any
> damage to the war-industrial factory. This is
> extremely important for the commercial class.
> ...Give it your personal attention.[46]

A study conducted in November 1916 by the Assistant Minister of War, N.P. Garin, found that 340
enterprises located in the central industrial
region signed contracts with the Moscow VPK. Of
these, 87 employed more than 1000 workers, 32 had
from 500 to 1000, 119 from 100 to 500, 47 from 50
to 100, and 55 enterprises contained less than 50
workers.[47] Based on the Moscow committee's own
lists which appeared in every issue of its
Izvestiia and another compiled by the Ministry of
Trade and Industry's factory inspectors, a similar
analysis can be produced for enterprises in Moscow
province.[48] Between 27 August 1915 and 30
September 1916, the Moscow committee distributed
orders to 142 such concerns (see Table 5.5).

TABLE 5.5 *Contractors of Moscow War-Industries
Committee located in Moscow province, 1915-16*

Branch of industry	Number of enterprises			
Metalworks	63			
Electrical	8		500 or more workers	33
Textile	25		100-500 workers	46
Leather	15	of which:	50-100 workers	32
Wood processing	8		less than 50 workers	31
Other	23			

The most frequent contractor was also one of the
largest heavy industrial enterprises of the area.
The Moscow Metalworks Company (formerly Goujon)
concluded at least twelve separate agreements with
the Moscow VPK to provide nuts, nails, scrap iron,
barbed wire and the like. These contracts were
facilitated by the active participation in the

committee of Jules Goujon, the director of the firm. Other heavy industrial firms which employed over 1000 workers and which concluded contracts with the Moscow committee were 'Iznoskov', Bromley Brothers, Dobrov and Nabgolts, the Moscow Carriage Works, and the Moscow War-Industry Company. Significantly, the Kolomna Machine Works, closely tied to the Petrograd International Commercial Bank and as such Petrograd's hostage in Moscow province, did not appear on the ledger of orders distributed by the Moscow VPK.

As far as shell orders were concerned, the committee's lists reveal that the majority did not go to the largest firms, but rather to those in the second category, with from 100 to 500 workers. Many had no previous war-industrial experience and depended on the committee to furnish drafts, industrial materials, and advances. Such enterprises were K. Fabergé, P. Khlebnikov and Sons, and P.I. Olovianishnikov.[49] Previously engaged in the manufacture of jewellery and church wares, these enterprises accepted numerous orders from the central and Moscow committees for grenades, shells and their parts, and concomitantly expanded their facilities. According to investigations conducted by the Moscow VPK, Fabergé employed 411 workers in May 1916 whereas before the war it contained only 181. Corresponding figures for Khlebnikov were 820 and 300-400.[50]

Moscow's pattern was duplicated elsewhere. The statistical bureau of the Special Council of Defence recorded 120 enterprises fulfilling orders for the Kiev VPK. The majority had not previously provided war material.[51] These newly converted firms and others mobilized by committees throughout the empire were the backbone of the movement. The decision to contract with the VPKs appeared to make good sense in terms of business as well as patriotism. If conversion involved risk, an enterprise incurred almost certain disadvantages by remaining outside war-industrial production. In return for the promise to fulfil the committee's order, an entrepreneur received assistance in obtaining fuel, raw materials, machinery and skilled workers. Given transport difficulties and the fact that ordinarily most of the firms would have been ignored by the military departments, the committee's *quid pro quo* for the production of war

material was not to be lightly dismissed. All too
often, however, the committees failed to deliver
what they had promised, and, consequently, so did
the enterprises.

5.3 THE WAR-INDUSTRIES COMMITTEES' ENTERPRISES
AND THE DEVELOPMENT OF NEW INDUSTRIES

It was to supplement production by their contractors as well as to supply them with the means of
production that the committees founded their own
enterprises. The extent to which district and
local VPKs engaged in this endeavour depended on
their financial resources, their complement of
technical and managerial personnel, and the availability of unutilized industrial facilities.
Enterprises were obtained by the committees
through purchase, by lease, or as sequestrated or
evacuated property to be returned to the original
owners at an appropriate date. According to the
committees' statutes, they could only accept
orders for their enterprises which did not include
a margin of profit.[52]

The Samara district VPK was the first to operate
its own enterprises and remained one of the most
heavily committed to this form of production.
Already in August 1915 it acquired two factories
employing a total of 750 workers and accepted
orders for grenades and shells. In December, the
'Muravei' factory, employing 552 workers, was
purchased outright by the committee for 741,000
rubles.[53] That the purchase and operation of
these factories was a financial burden to the
committee is suggested by a telegram which its
chairman addressed to the central committee in
February 1916. 'Situation desperate', it said. 'If
no money forthcoming, committee wil be forced to
cease activity.'[54] The Kiev committee, generously endowed by its chairman and containing a
large number of technical personnel, operated 10
facilities, more than any other committee. The
Moscow VPK ran 9, while local committees within
its area of operations supported an additional 21
enterprises.[55]

By 1917, 59 committees (42 local, 17 district,
and the central committee) ran 120 enterprises.
They included: 45 producing shells, artillery

parts and machines; 18 fulfilling orders for field
kitchen units, carts and horseshoes; 31 involved
in the production of footwear, saddles and hides;
15 chemicals factories; 5 producing medical
supplies; 4 food processing factories; 3 electrical goods factories; and 1 producing gas
masks.[56] The enterprises ranged in size from
small shops employing 5 or 10 people to large-
scale plants such as the Baku committee's shell
factory which employed as many as 920 workers, the
'Sotrudnik' Machine Works of the Saratov committee
which employed 530, and the Samara committees'
grenade assembly facility. It would appear that
the latter hardly deserve the epithet of 'cottage
industries' which has been used to describe the
VPKs' enterprises.[57]

The enterprises were singularly adapted to
wartime. At least three facilities – a boot
factory in Krasnoiarsk, a saltpetre laboratory
belonging to the Central Asian VPK, and a grenade
assembly unit administered by the Rostov-on-Don
committee – relied on prisoner-of-war labour.
Several were unique or one of few of their type in
Russia. A laboratory of the Baku committee
produced toluene, an essential ingredient in high
explosives; a factory in Kazan', managed by the
district committee, experimented with phenolsali-
cylate with an aim toward mass production; the
Kiev committee supported optical instrument,
pharmaceutical, iodine and chloroform laborator-
ies; and the central committee's medical supply
section opened the first facility in Russia for
the production of clinical thermometers.[58] The
most extensive undertaking along these lines was
the central committee's 'Respirator' enterprise,
one of two in Russia to produce gas masks. The
enterprise consisted of six factories, all located
in the Petrograd area, and employed a total of
3000 workers.[59] At its peak, 'Respirator' turned
out from 25,000 to 30,000 gas masks per day, and
by August 1916 it had manufactured over 1
million.[60] On a smaller scale, the Moscow
committee's chemical section under Professor
Chichibabin operated a laboratory for the
production of phosgene, the poison gas.

In the pre-war era, Russia was particularly
deficient in the production of chemicals and
precision instruments, relying on foreign imports,

primarily from Germany. The war created the need for the domestic development of these industries and the VPKs, having mobilized a significant section of the Russian scientific and technical community, played an important role in fostering it. Apart from operating their own enterprises, the committees served as a forum for the discussion of new products and methods of production.

Not infrequently, proposals emanating from such discussions conflicted with those developed by parallel government bodies. Such was the case with a plan drawn up in July 1915 by the TsVPK's chemical section for the construction of 2000 coking ovens in the Donets Basin. As the Explosives Commission, formed in February under the auspices of the GAU, already had its own plan for the expansion of coking facilities, its chairman, General Ipat'ev, considered the committee's 'disruptive'. He argued for its rejection before the Preparatory Commission of the Special Council of Defence and succeeded in winning a majority.[61] Such, as we shall see, was also the pattern in the regulation of industrial relations, production and distribution. It resulted from the failure of the government adequately to integrate private enterprise and the scientific knowledge it could mobilize. In this sense the self-mobilization of industry worked at cross-purposes with the state's, even if unintentionally so.

On the other hand, in the area of technical research and its application to the war effort, the committees were practically alone. The central and Moscow committees each sustained inventions sections, the purposes of which were to encourage technical research and to receive and pass judgement on the feasibility and relevance of proposals. The central committee's section boasted 73 professors, engineers, artillery and other military experts.[62] The Moscow section was under Professor N.E. Zhukovskii, 'the father of Russian aviation'. Within its first 4 months of operation, the section had received 206 proposals of which 11 were considered worthy of further study. A month later, the number of proposals reached 443. Of the 936 proposals received by September 1916, 317 had to do with artillery, 162 with machines, 103 concerned aviation, 109 chemicals, 35 bridge construction, 44 electrical devices, 115 combinations of the above, and 78 were not categorized.[63] The final report of the section in October 1917 listed *inter alia* the following inventions as having been 'constructively developed' by the section's staff: a device to reduce the recoil of cannons, a relay for telephones with voice frequency signalling, an instrument to improve aerial machine-gunning, improvements in wide-angle photography, and a pocket goniometer.[64]

The Moscow committee popularized as well as organized work of this nature. It sponsored exhibitions and published numerous articles promoting the production of pharmaceuticals, chemicals and aeronautical instruments. With reference to the application of refrigeration methods, one article explained 'how England and its army are fed', 'the supply of France with meat', 'how Germany is fed', and 'what Russia can do'.[65] In October 1916, after several months of negotiations with the Ministry of Internal Affairs, the committee hosted the first and only All-Russian Congress of Inventions to be held in Tsarist Russia.[66]

The culmination of the Moscow committee's work in promoting industrial development and at the same time a new departure was represented by the journal, *The Productive Forces of Russia*, the first number of which appeared in November 1916. This was more than a technical journal, though it was that too. Devoted to 'all questions related to future economic construction', it sought to marshal support for private investment in Russian industry, the development of Russia's hinterland, and other projects intended for the post-war period. As such, it falls outside the discussion of the committees' war-industrial endeavours and will be taken up in the context of efforts to extend their life beyond the war.

5.4 THE FULFILMENT OF ORDERS

It is ironic that one of the first tasks which the TsVPK performed for the Special Council was to inform it of the late fulfilment of orders by the GAU's major contractors. Finding that the five largest armament producers in Russia had fulfilled only three-quarters of their orders due by July 1915, the committee made several recommendations for the improvement of purchasing and transport operations.[67] Within a few months, the situation was somewhat reversed: the VPKs were the object of the Special Council of Defence's investigations, and they revealed a much poorer rate of fulfilment.

The fulfilment of the VPKs' orders was the ultimate test of the organization's worth to the army. As a report by the central committee's accounts section emphasized,

> the view that the sum of orders which the central committee distributes serves as a criterion of its activity is completely mistaken. Quite the contrary is true. The distribution itself represents a relatively simple operation; far more complex is the supervision of fulfilment, supplying the customer with raw materials and other facilitating operations.[68]

It also sheds light on the degree to which the committees were dependent on the government, and the efforts of civil and military officials to exploit this dependence to the committees' disadvantage.

As early as September 1915, the central committee was informed that the organization was not properly processing its orders. Professor Zernov, briefing the sixth session, conceded that the mechanical section operated 'without a general plan'. Baron Maidel warned that the hopes placed in the committees, hopes which the organization itself inspired by its grandiose schemes, 'are considerably exaggerated'.[69] There was disquiet in other quarters as well. General Manikovskii, never a supporter of the committees, pointed out in an interview in October 1915 that the front-line units had not received a single shell from them.[70] The extreme right seized upon Manikovskii's remark. A monarchist congress held in Nizhnii Novgorod in November resolved that the VPKs should be closed because 'they are useless and only give society the impression that they are helping to carry on the war'.[71]

In response to such claims and threats,

partisans of the committees sought to defend their record. Professor M.A. Sirinov, who wrote many leading articles for the central committee's *Izvestiia*, pointed to the difficulties involved in converting enterprises to munitions production and asserted that despite them the committees' first orders had been fulfilled on time.[72] What delays there were, another publicist explained in terms of 'infantile disorders' to which every young organization was prone. These explanations seemed to satisfy the press, evoking a ringing endorsement of the committees even from the right-wing *Novoe Vremia*.[73]

The first investigation of the committees' activities conducted by the Supervisory Commission of the Special Council of Defence told another story. It revealed that as of December 1915 not one order from the GAU had been fulfilled according to schedule. Instead of an expected 2045 bomb-launchers, the TsVPK provided only 1309. Of 699,500 shells, only 248,000 had arrived. Grenade deliveries were far worse. The army expected 2,257,000 but received from the committees only 122,000. Not one mortar shell had been delivered although over 80,000 were due. The commission, which numbered among its members Konovalov and A.I. Shingarev, was more than tolerant of the committees' failings and merely recommended that they 'devote their attention to the greatest possible fulfilment of the Ministry of War's orders'.[74]

The fulfilment of orders continued in a desultory fashion throughout December. The almost geometric increase in deliveries thereafter suggests that a certain period was required to establish patterns of communication with enterprises and arrange for the delivery of materials to them. This lead time was the inevitable but none the less unfortunate price which Russia paid for the late arrival of its provincial industrial bourgeoisie to munitions production. That the government might have organized mass production in a shorter space of time without the 'interference' of the public organizations is a moot and rather doubtful point. Even the Vankov Organization, that highly efficient, much praised network of partially converted enterprises presided over by a team of military technicians, did not begin to deliver large quantities of shells until late 1915 or early 1916.[75]

In February, the TsVPK delivered as many of its most frequently ordered articles as it had up to 22 January. Deliveries in March continued this trend as indicated in Table 5.6.[76]

TABLE 5.6　*Deliveries by Central War-Industries Committee of selected articles, 1916*

Articles	Deliveries		
	22 Jan.	22 Feb.	22 Mar.
1　9-cm bomb-launchers	1,736	2,908	4,031
2　9-cm shells	175,497	373,411	690,648
3　Hand grenades (1914 type)	19,978	159,460	539,043
4　Hand grenades (German type)	256,000	500,000	991,937
5　58-mm mortars	516	605	725
6　58-mm mortar shells	133	860	8,258

The activity of provincial committees presents a variegated picture, but one that shows general improvement during the first half of 1916. The reports of district and local committees to the second congress provide the earliest and most depressing record. The Revel VPK reported only 450,000 rubles worth of orders fulfilled of a total of 3.14 million distributed. The district committee of the Caucasus had a worse record. It fulfilled orders amounting to a mere 86,000 rubles or 7.1 per cent of the amount distributed. This was at least better than the Viatka committee which had not fulfilled anything though it had received orders valued at 1.1 million rubles.[77] While it is impossible to determine from these reports the anticipated delivery schedule, a survey conducted by the Special Council of Defence's provincial units, the factory councils (*zavodskie soveshchaniia*), reveals widespread tardiness. Of 13 orders for the GAU distributed to Siberian committees, only 1 for 20,000 shells was fulfilled on schedule by 20 February 1916.[78]

Higher rates of fulfilment are evident after the second congress. The Moscow committee, which had fulfilled only 20 per cent of its distributed orders by February, could claim 30 per cent 2 months later. Thirty per cent was cited by the Saratov committee in a report of May 1916. The Rostov-on-Don district VPK listed 3.7 million rubles in orders fulfilled of 10.9 million distributed by May, and several local committees in its area were singled out for praise by the factory council of the Caucasus.[79]

The increase in deliveries from provincial committees is reflected in the general rise of orders fulfilled by the TsVPK. At its nineteenth session on 28 March 1916, Maidel remarked on the 'massive production' but added that fulfilment was still behind schedule. Professor Sirinov announced in mid-May that 25 per cent of the GAU's orders and 50 per cent of the GIU's (excluding horseshoes and nails) had been fulfilled.[80] By the middle of the year, the TsVPK delivered 88 per cent of all bomb-launchers on order, 58 per cent of their shells, 76 per cent of all cast-iron (German type) hand grenades, and 53 per cent of all mortars. On the other hand, only 16 per cent of all tin-plated (1914 type) hand grenades on order and 12 per cent of all mortar shells had been delivered.[81] The figure of 70 per cent supplied by Professor Savvin to the bureau of the central committee in December 1916 certainly exaggerates the rate of fulfilment by that month.[82] The estimation of slightly more than one-third for orders distributed up to the February revolution is probably closer to actuality.[83]

The early non-fulfilment of orders and the persistence of low rates for certain articles became an issue on which the fate of the organization turned. The yawning gap between orders accepted and fulfilled threatened to swallow up the committees on several occasions. The conditions of fulfilment were such as to provide both the committees' defenders and their detractors with arsenals of arguments which, hurled at each other, scored many hits. The former armed themselves with the righteousness of public activists (*obshchestvennye deiately*) who were thwarted by an unsympathetic and *ipso facto* irresponsible bureaucracy. The bureaucrats cast

themselves as the guardians of the state treasury
which power-seeking *poseurs* had dented by their
unceasing demands for orders and advances on them.
In the fray, which had strong political overtones
and owed much of its vituperativeness to the
resolutions passed by the committees' second
congress, the bureaucrats received the support of
right-wing elements in the Duma and the press.

Internal disorder constituted the most powerful
indictment against the committees. The unplanned
nature of distribution did not bode well for the
fulfilment of orders. Professor Grinevetskii, the
chairman of the Zemgor's engineering section, was
struck by the irrationality of the TsVPK's policy.
Orders for the relatively simple 3-inch shell were
granted to firms with experience in munitions
production, while smaller enterprises received
orders for the more technically exacting 42-line,
48-line and 6-inch shells. This the professor
considered 'a mistake of colossal significance'.[84] The high rate of cancellation of
orders is further evidence of the poor judgement
exercised by the central committee. While the
Ekaterinoslav VPK annulled orders valued at
244,500 rubles or less than 5 per cent of what it
distributed, the central committee cancelled over
8 million rubles worth of orders to enterprises in
Ekaterinoslav province.[85] Such was also the fate
of 1.5 million hand grenades ordered from the
Pitoev factory in Batumi. As far as the 48-line
and 6-inch shells were concerned, the committee
had annulled 288,000 of the former and 275,000 of
the latter by May 1916.[86]

The second charge levelled against the
committees concerned the allegedly inflated level
of their prices. This was a sensitive issue
because both the VPKs and the Zemgor advertised
themselves in terms of the relative cheapness of
their orders. It was, so they claimed, standard
practice to conclude agreements with enterprises
at prices below those fixed by the military
departments. Since the amount of material
delivered to the army's warehouses was expressed
in terms of the actual and not recommended prices,
this would also put the organizations' rates of
fulfilment in a better light.

Which claim was true depended on the source as
well as the particular item in question. According

to Konovalov, as of the beginning of December
1915, the committees had saved the state 20
million rubles. Twenty-six million rubles was the
figure cited by Zhukovskii to a special meeting of
representatives from district committees in May
1916.[87] The Moscow committee alone claimed to
have saved the state nearly 18 million rubles by
placing orders below recommended prices.[88] The
Council of Ministers, on the other hand, correlated the VPKs' activity with the general rise in
prices as early as November 1915:

> This circumstance demands in the opinion of the
> council special attention since the cost of
> orders transferred by the War-Industries
> Committees does not appear to correspond to the
> actual cost of preparing the article and there-
> fore raises the cost of the state's administrat-
> ive activity to the extreme.[89]

For all the weight that Manikovskii and certain
Soviet historians have given to the ministers'
assertion, it rested on incomplete and misleading
information.[90] The inflated prices cited by the
ministers had little to do with the VPKs which
accepted orders for only two of the six items they
mentioned. One of these was the 3-inch cannon, the
production of which the central committee failed
to organize. The other, the 3-inch shell, owed its
high price as much to Major-General Vankov, the
principal expediter, as it did to the VPKs. More
concerned with finding scapegoats than the real
causes of inflation, the ministers all too easily
mistook cause for coincidence. K.N. Tarnovskii in
his authoritative study of the metallurgical
industry in Russia has noted that from May 1915
the world price of all metals 'jumped sharply'.
This, plus speculation by Russian banks and steep
price increases by Prodamet, it is suggested, had
a greater effect than the pricing policies of the
War-Industries Committees.[91]

However misguided the aim of the ministers, they
hit their target. Of 26 articles produced by the
Zemgor and the VPKs, the former charged more for
only 7 and its prices averaged 6 per cent
less.[92] Moreover, the committees concluded
contracts for certain items at prices well above
those established by the military departments. The

price of the explosive appurtenance to the 58-mm
mortar shell was set by the GAU at 6.70 rubles but
produced by the VPKs' contractors for 9.65 rubles.
While this excess cost was compensated by the
central committee's contracts for the remaining
parts of the shell, those concluded by the Moscow
committee were still higher than the GAU's price
by 5.50 rubles.[93] The committees did even less
justice to the Quartermaster Unit. The military
stores section of the central committee admitted
that 'prices by which orders were delivered,
despite all efforts by the section, could not
always be particularly low, nor could the period
of production be short'. The section referred to
its orders for cavalry equipment which, at 155
rubles, were 35 to 45 rubles above the 'normal
price' though still below the department's. It did
not cite its prices for horseshoe nails. These
were ordered by the committees for 50 rubles per
1000 or approximately 18 rubles above the GIU's
recommended batch price.[94]

Instances of genuine corruption involving the
committees were assiduously recorded by the
police. Corruption took two forms: profiteering by
brokers, and illegal or unethical practices of
individual VPKs. The first falsified the notion
that the organization constituted a reliable
alternative to unscrupulous middlemen. The
Ekaterinoslav VPK, cited above for the perspi-
cacity of its distribution policy, placed orders
with 65 firms, but 10 of these had been contracted
via intermediaries. One notarized agent, M.B.
Katsnel'son, charged a commission amounting to 30
per cent of the orders' value. The VPK's own
control commission exonerated the organization but
the case went before a military judicial
tribunal.[95] The central committee's reliance on
Russian agents to facilitate the purchase of metal
and machinery on the English market was condemned
as wasteful in a report from London to the Special
Council of Defence. One agent in particular, V.V.
Behr, was described as 'clever and adroit at
playing the great gentleman, but according to the
English [is] one of the biggest sharks fattened by
the war'.[96]

Given that the committees consisted of thousands
of businessmen who were themselves performing an
intermediary function, malfeasance by individual

committees or committee members was inevitable. The multitude of such cases must, in any event, raise doubts about the public-spiritedness of a sizeable portion of the organization. The most common abuse was the channelling of profitable orders to enterprises in which one or more of a committee's members had a special interest. One such instance, which the Moscow committee's investigator politely described as a 'misunderstanding', concerned eight factory owners who had established a branch of the Nizhnii Novgorod VPK in Pavlovo for the sole purpose of receiving orders and distributing them among themselves. This was not as objectionable as the fact that they charged higher prices than the local *artel*.[97] In Novocherkassk, a member of the VPK worked in reverse. He first negotiated several 'good' orders with a local enterprise and then, having signed the contracts in the name of the committee, purchased the enterprise.[98] The least reputable of all the committees, however, was the Arkhangel'sk district committee, which was eventually reorganized under the supervision of A.A. Bublikov. Its violations of the public trust filled eight pages of a report sent to Guchkov on the eve of the reorganization.[99]

Apart from the committees' own shortcomings, problems inherent in the production of military supplies plagued the organization and its contractors, especially those new to war-industrial production. Accustomed to dealing with a few experienced suppliers, the GAU's officials demanded too much from the novices. The army's needs were such that long-term orders were granted grudgingly. Yet even for short-term orders the committees often had to wait months to receive the technical specifications. Blueprints and prototypes were on occasion denied to them and those provided were often technically inadequate. It was the committees' own draftsmen who defined the margin of error in the production of the 3-inch shell. The GAU itself was not sure of the specifications for the 6-inch and 48-line shells, and twice altered them.[100]

Enterprises which proceeded with production despite unclear instructions risked the wrath of the GAU's inspectors. According to a delegate from Moscow to the second congress, these guardians of technical exactitude were 'largely inexperienced and therefore require a greater degree of precision than is in fact necessary'.[101] Several factory owners reported to the Moscow committee instances of harassment by inspectors.[102] Not until November 1916, after several incidents embarrassing to the GAU, did the committees obtain the right to assign their own inspectors to factories fulfilling their orders.[103]

The transport system imposed yet another obstacle to the fulfilment of orders on time. The list of priorities according to which deliveries of materials and fuel were made favoured enterprises in direct contact with the military departments. Those contracting with the VPKs received lower designations.[104] Since, as the Special Council for Transport admitted, even top priority material was frequently delivered behind schedule, it comes as no surprise to learn that those further down the list waited months and sometimes in vain.[105] A report filed in May 1916 by the TsVPK showed that the Revel committee had been waiting since the previous October for 60,000 puds of pig iron, the Tver VPK even longer for tin-plated cases, and the Odessa VPK since November for 249 coal wagons.[106] In other reports submitted by provincial committees, the term 'disorder of transport' was often cited in connection with the late fulfilment of orders.

Finally, financial restrictions placed on the committees constituted a most serious hindrance to their economic activities and ultimately to the fulfilment of orders. This was not due to the whim of a few military officials or the application of discriminatory regulations, but was an explicit policy of the government in spite of its own rules. This policy, as will be demonstrated, effectively sabotaged the development of the new war-industrial network which the committees tried to forge.

The committees unwittingly contributed to their own financial difficulties. Expecting a large flow of orders, delegates to the first congress approved the provision for the deduction of 1 per cent of their cost to cover organizational expenses. This, it was felt, would give the committees more independence than the Unions which relied on state subsidies. The amount obtained from this source and voluntary contributions was, however, insufficient. Many provincial committees experienced severe financial strains. The second congress sought to rectify the error of the first by proposing an increase in the percentage, but to no avail. An inspector from the State Comptroller's office considered the proposal 'completely unacceptable'.[107]

The committees undertook a number of commercial operations to augment their meagre income. Thus, the central committee sold to Russian enterprises foreign machinery and steel at prices 10 and even 20 per cent above what it had paid.[108] The export of potash over which the committee exercised a monopoly was executed at a 2 per cent commission. Several committees sold scrap metal from their stocks collected at the front. According to the Moscow committee's accountant, this constituted its main source of income, amounting to 2.58 million of a total of 5.57 million rubles by May 1918.[109] Finally, orders for the front-line units were fulfilled at a 3 per cent commission. Yet, these ventures had one obvious drawback. They were, by and large, at the expense of the very enterprises whose production the committees tried to stimulate.

Vital to the committees' network of contracting enterprises was working capital to purchase machines, raw material, fuel and skilled labour power. As already mentioned, it was the committees' promise to attract the necessary capital which induced many entrepreneurs to accept their orders. Generating no capital of its own, the organization directed its efforts towards the extraction of advances on orders from the state. According to rules confirmed by the Special Council of Defence on 16 November 1915, both the VPKs and the Zemgor were entitled to receive 50 per cent of the cost of orders as advances.[110] The failure of the military departments to abide by these rules, especially in 1915, led to numerous protests on the part of the committees.

The VPKs' campaign to receive advances, which became a *cause célèbre* for the public organizations, actually commenced before the issuance of the rules. In October, the central committee's *Izvestiia* cited the amount distributed by the departments as 'quite insufficient'. By December,

Zhukovskii was calling the departments to account for their apparent neglect. While orders distributed to the committee amounted to 185 million rubles, declared Zhukovskii, it had actually received only 16 million rubles.[111] For the purchase of machinery, advances amounted to 29.7 per cent of orders, but for shell and grenade production, which entailed the acquisition of metals and coal fuel, only 6.1 per cent of the value of orders was provided. Advances on stores came to a mere 2.7 per cent.[112]

By August 1916, the last month for which such figures are available, advances amounted to 18.3 per cent according to the accounting section or 31 per cent according to a memorandum addressed to the Ministry of War.[113] Although the minister, Shuvaev, ordered the payment of additional sums and promised to adhere to the rules of 16 November in the future – indeed extending them to allow for advances of up to 75 per cent of an order's value – such efforts could not make up for the lost time.[114] Furthermore, it is not at all clear to what extent Shuvaev's instructions and promises were carried out. The central committee's accounts for 26 September 1916 show 91.86 million rubles in advances, only 3 million rubles more than the account of 1 August.

Such parsimony was not entirely motivated by concern for the state treasury. The normal procedure for granting credits cost enterprises anywhere from 3.5 to 15 per cent of the advance, and this charge indirectly led to an increase in the cost of orders.[115] The procedure was based on the Council of Ministers' resolution of 23 July 1914 according to which enterprises were entitled to receive up to 65 per cent of the order as an advance if they could furnish from commercial banks 'letters of guarantee' to certify their credit worthiness. The banks would grant the letters at their own discretion and for a fixed percentage of the sum received by the enterprise as an advance.[116] The Russian-Asian Bank, for instance, provided such letters to Putilov Works, Nevskii Shipbuilding, and other favoured enterprises at rates from 3.5 to 6 per cent.[117]

The TsVPK as a guarantor was cheaper. It accepted interest bearing securities amounting to only 2 per cent of the advances passed on to its contractors.[118] According to regulations approved by the Special Council of Defence in November 1915, enterprises certified by the committee were exempt from the earlier provision requiring a letter of guarantee.[119] Nevertheless, the committee was used sparingly by the departments. Between 15 September 1915 and 1 August 1916, it was asked to investigate only 219 firms, of which 186 received the committee's certification.[120] Despite pressure from within the Ministry of War to do away with the banks' letters, numerous litigations in late 1916 suggest that the banks continued to extract sizeable sums for them.[121]

Such competition between the War-Industries Committees and financial-war-industrial conglomerates was compounded by the search for the materials of production. For even those enterprises which received advances or had otherwise amassed working capital, there remained this final step before production could proceed. The large Petrograd-based enterprises relied on their patron banks to enable them to purchase outright or buy into coal, iron ore and steel companies, a process of vertical integration that had begun during the fuel 'famines' before the war. Thus, for example, the acquisition of the Novorossiisk Ironworks Company by the Russian Company for the Production of Shells in April 1916 was really a matter of internal bookkeeping for the Russian-Asian Bank.[122]

By facilitating the regular and relatively cheap flow of fuel and raw materials, such integrations enabled client enterprises to expand their lucrative shell sections. While the Putilov Works was becoming embroiled in financial and labour disputes – which would result in its sequestration by the state in February 1916 – and the VPKs were finding it increasingly difficult to supply their contracting enterprises, those affiliated with the Petrograd banks and/or Vankov surged ahead in munitions production and the profits derived therefrom. Of an estimated 39 million shells produced in Russia during the war, 28.3 million, or over 70 per cent, were turned out in 1916.[123] The majority came from the same dozen or so firms on which the state had depended before the war. It was thanks to such progress in shell output, plus similar advances in chemicals, artillery and small arms production that the munitions shortage of 1914-15 was largely overcome.

But in the meantime, munitions had ceased to be the sole or even main object of dispute in the politics of industrial mobilization. The alliance between the state and Petrograd financial-industrial circles, strained by the munitions shortage but reinforced by the challenge which the VPKs posed to both, virtually ensured that every aspect of the war economy would come under the committees' scrutiny and be found wanting. Hence, the regulation of the economy became yet another battleground between the state and the War-Industries Committees.

5.5 CONCLUSIONS

The VPKs obtained a substantial sum of orders from the military departments - approximately 400 million rubles by the February Revolution - and supplemented this source with orders from military units, state factories and other public organiz-ations. This quantity, however, was not so large as originally expected, and hardly warranted the extensive network of committees which sprouted on the Russian soil immediately following the Associ-ation of Industry and Trade's ninth congress. Hence, while some committees 'burned with a fervent desire to immediately receive orders' and another claimed that 'to satisfy the general desire to work for defence it is necessary to obtain orders and more orders', they were to be keenly disappointed.[124] The failure of the central committee to meet the expectations of district and local committee members, the fact that over half the sum of orders accepted by the central committee was distributed to enterprises without the participation of the provincial committees, made many of them redundant.

The main reasons for the failure of the TsVPK to garner a greater share of state orders were the unwillingness of the largest firms to forgo their direct connections with the departments and the departments' refusal to permit a wedge between themselves and the enterprises. Nevertheless, the committees did represent an important conduit for

orders of certain items which either had not been
manufactured in Russia at all in the pre-war era
(3-inch high-explosive shells, bomb-launchers, gas
masks) or whose production by large metalworks
factories (grenades, mortars) had diverted these
enterprises from more complicated endeavours
requiring greater outlays of capital. Among the
committees' contractors and their own enterprises,
middle-range and small plants predominated, though
large firms employing upwards of 1000 workers also
accepted orders, especially from the Moscow VPK.
The Moscow committee also played an indirect but
no less important role in textile production for
the army, assuming duties formerly performed by
the Moscow Exchange Committee.

These and still other services – statistical
surveys, the promotion of war-related inventions
and industries, assistance in the evacuation of
plants from areas threatened with occupation –
earned the committees a place in the Russian war-
industrial complex. But, to repeat, it was not the
sort of place which many committee members had
been led to expect. It was, as General A.Z.
Myshlaevskii, the chairman of the state's metallurgical committee, remarked, more in the order of
'a salutary corrective and supplement to the
existing system of supply to the army'.[125]

In terms of their primary responsibility, the
fulfilment of orders, the committees themselves
often required correction. Deliveries were behind
schedule; the cost of production was higher than
anticipated; graft was widespread. Yet, it is
difficult to hold the committees entirely responsible for such shortcomings. Badly underequipped,
the army needed supplies sooner and on a much
larger scale than Russian industry could provide.
The backbone of the committees' network, medium-
range enterprises often converted from market
production, were especially hard pressed to fulfil
short-term orders. Such enterprises required not
only time, but money. Their fixed capital could
not be supplemented or transformed unless working
capital was available.

The VPKs' ill-defined and often strained
relationship with the military departments and
ultimately with the government precluded large
disbursements of credit. The failure of the
government to adhere to its own rules with respect

to advances can be attributed to its reluctance to give the committees a financial base. Such a policy was consistent with its reliance on commercial banks to furnish letters of guarantee, and the limited use it made of the central committee's certifying operations.

6 The War-Industries Committees and the Regulation of the War Economy

The participation of the War-Industries Committees in the regulation of the war economy was, to some extent, the natural outgrowth of their acceptance of state orders. The production of munitions inspired a concern for metals production, distribution and the costs involved. Military stores suppliers themselves required an adequate supply of raw fibres, of wood and of animal hides. The determination of the demands of industry and commerce was closely related to the resolution of transport difficulties, and this with fuel supply. Machinery and metals which Russia produced in insufficient quantities had to be obtained from abroad and this brought the VPKs into contact with the international market.

But the committees' regulatory activities were never intended to be confined to facilitating their fulfilment of orders. Their leaders had pretensions of filling the void between the requirements of war and what they regarded as the government's dim awareness of them. To be sure, the state had not been idle in the first year of the war. Numerous decrees restricting the free flow of workers and goods had been passed, several special commissions had been created, and an even larger number of proposed measures were under investigation. Yet, the system was not functioning in any coherent manner, or rather there was no system. Areas of competence were only vaguely defined among civil authorities and between them and the military. In the area of transport, for example, 6 different agencies, 4 civil and 2 military, could issue special orders to commandeer supplies.[1] Inevitably, one commission cancelled out the actions of another. This was both

inadvertent and illustrative of genuine conflicts
between ministers. On several occasions, the War
Ministry advanced proposals for placing state
enterprises on a special condition tantamount to
militarization, but the Council of Ministers
invariably rejected them on the advice of the
Minister of Internal Affairs.[2] Altogether,
despite the creation of a virtually parallel
bureaucracy in the form of the Special Councils,
there was no agency which coordinated plans or,
for that matter, developed them.

The industrial bourgeoisie was itself ambivalent
about the role of the state in the regulation of
the economy. Having fought for years to be free of
the state's 'guardianship' (*opeka*), even the
Association of Industry and Trade recognized the
need for strict regulation during the war. But
whether the bureaucracy, accustomed to only
internal accountability, could manage the task was
considered doubtful.[3] The market had ceased to
be an indicator, if in the Russian context it ever
was, of what should be produced and how much it
should cost. The war strengthened tendencies
towards a command economy but had not produced a
commander. 'It is the tragedy of Russian society',
remarked the Kadet politician, N.I. Astrov, to a
conference on the economic situation in January
1916, 'that while we do not believe in those who
hold power, we at the same time cannot but
acknowledge that only powerful authority can save
the situation.'[4]

Tragic it may have been, but the situation also
provided the opportunity for the forces claiming
to represent the 'public' to obtain a powerful
voice in the administration of the war economy.
The participation of the VPKs in economic
administration was a complex and contradictory
phenomenon. They became at once the repository of
quasi-official responsibility and the enemy of
state control over industry. They clashed with
monopolistic interests in some areas but came to
their defence in others. In the politically
charged atmosphere, the means for achieving a well
ordered system of production and distribution
became political issues in themselves. Some of
these issues will be explored in detail.

6.1 METALS AND FOREIGN PURCHASES

During the war, Russia possessed two major metallurgical centres: the Urals, some of whose enterprises dated back to the eighteenth century, and the Donets Basin of South Russia developed in the last two decades of the nineteenth century primarily by French and Belgian capital. Siberia's rich deposits of ore and the potential for developing them was a dream, and only the most hardy or foolhardy considered investing in it. The Dombrowa Basin, which had served Poland's industrial needs, was occupied in the first weeks of war and was unavailable for Russian exploitation. Other areas (the Moscow region, the Volga and Baltic districts) accounted for less than a sixth of Russia's total output.

Both major centres engaged in three-step production of ferrous metals, which is to say that the bulk of pig iron and semi-finished iron and steel was locally consumed. Both lent themselves to syndicalized distribution. The Urals ferrous metals syndicate, 'Krovlia', organized the sale of roofing iron of which that area supplied nearly three-quarters of Russia's needs.[5] More significant, especially with respect to the war effort, was Prodamet, an association of twenty of the largest enterprises in South Russia.

The policy of the VPKs with respect to ferrous metal was dictated by their relationship to the metal syndicates and, particularly, Prodamet. As a show of support for the new organization, representatives from eleven enterprises belonging to Prodamet agreed in June 1915 to establish three special 'information organs' under the auspices of the TsVPK's metallurgical section.[6] One was to cover the southern and central (Moscow) areas, a second to apply to the Urals, and a third to deal with all questions related to non-ferrous metals. The metallurgical section, chaired at the time by N.S. Avdakov, promptly accepted into its ranks several prominent directors of the syndicate. They included its long-standing president, P.G. Darcy, V.A. Vvedenskii, D.B. Vurgaft, I.I. Efron and S.A. Iablonskii.[7]

While Prodamet was consolidating its position in the central committee, members of other committees were growing wary of the potential conflict of interest represented by the 'information organs'. The first congress expressly rejected them and passed a resolution indirectly critical of Prodamet (see p. 67). Shortly after the congress, another plan emerged from within the metallurgical section. This called for the creation of a Bureau for the Distribution of Metals (hereafter the Metals Bureau). The Metals Bureau was intended to determine the requirements in ferrous metals of all factories and the state, assess the productive capacity of metallurgical enterprises, develop ways of increasing it, and cooperate in planning the distribution of orders for steel.[8] Representation on the new bureau was such as to establish a balance between Prodamet and the Urals metallurgical industry. The chairman of the bureau was N.N. Kutler, the head of the Association of Urals Mineowners as well as of the Urals Mining and Metallurgical VPK.[9] On the other hand, Prodamet's V.A. Vvedenskii replaced the deceased Avdakov as chairman of the metallurgical section.

The Metals Bureau in its 4 months of operations under the TsVPK fulfilled the first two of its tasks. It estimated that the military departments and the railways would require 177.5 million puds of metal for the period November 1915 to January 1917. Requirements for industrial enterprises were forecast as 63.8 million, bringing the total to 241.3 million puds. In terms of pig iron, this meant that 266 million puds would be needed for the 14 months or 230 million during 1916. Secondly, it projected an output of pig iron of precisely 230 million puds.[10]

But no sooner had the Metals Bureau submitted its projections, than it was superseded by another organ, the Metallurgical Committee, a state institution reporting to the Minister of War.[11] The Metals Bureau's sudden loss of stature caused much concern at the time. N.V. Nekrasov, a leading member of the Zemgor organization, claimed that the VPKs 'were compelled' to liquidate their Metals Bureau, but did not reveal who compelled them, how, or why.[12] At the committees' second congress, Baron Maidel, responding to P.I. Pal'chinskii's tirade against the central committee's apparent self-effacement, was somewhat more explicit. He informed the congress that 'we divested ourselves of our institutions because at crucial moments we could not overcome obstacles placed by the government'.[13] Still, no indication was given of what these obstacles were. Two possible explanations have been advanced by the Soviet historian, A.P. Pogrebinskii. He asserts that 'in opposition to the VPKs' Bureau for the Distribution of Metals, Polivanov set up a Metallurgical Committee'. Elsewhere, he writes of a 'vigorous struggle' between Prodamet and the TsVPK which resulted in a complete victory for the former and the consequential dissolution of the latter's Metals Bureau.[14]

As suggested by General A.Z. Myshlaevskii, chairman of the Metallurgical Committee, the initiative for the change actually arose from within the TsVPK.[15] In a memorandum to Polivanov dated 8 November 1915, the central committee proposed the creation of a metallurgical committee to fulfil, by the imposition of compulsory measures, the remaining tasks previously assigned to the Metals Bureau. Describing the proposed committee as a 'mixed state-public type of institution', the memorandum called for a chairman appointed by the Minister of War, one representative each from the ministries of the Navy, Trade and Industry, and Ways and Communications, and a total of twelve representatives from 'society' evenly divided among the two Unions, the TsVPK, and the Association of Industry and Trade.[16]

There appear to be two reasons why the central committee proposed the 'statization' of the regulation of metal. First, as the bureau's report observed, 'factories place their production under the control of the bureau on an exclusively voluntary basis which they are free to ignore at any moment'. As of November, it had handled orders for only 85,000 puds of iron and itself purchased an additional 160,000 puds.[17] This amounted to less than 1 per cent of the iron produced during the same period. Its petitions for distributing up to 650,000 puds of shaped steel per month and 400,000-500,000 puds of sheet iron to the oil industry were ignored by the government.[18] Secondly, it was no coincidence that simultaneously with the committee's proposal, the Ministry of Trade and Industry introduced one less favourable to 'society'. Its committee would also exercise executive authority but the chairman was to be an appointee of the minister, Shakhovskoi, and the ministerial plenipotentiaries were to be more numerous.[19]

To resolve the question of which proposal was more efficacious, Polivanov called a rare joint session of the Special Councils of Defence and Fuel on 25 November. Speaking in favour of Shakhovskoi's plan were two Octobrist deputies, A.D. Protopopov and B.I. Timiriazev, and the former Minister of Trade and Industry, S.I. Timashev. Zhukovskii, Kutler and Konovalov defended the TsVPK's proposal. Polivanov threw his crucial support behind the latter plan, and the session adopted it with some modifications.[20] On 1 January 1916 it began operations.

The Metallurgical Committee was a state institution but its establishment was a clear victory for Prodamet and the other metallurgical syndicates. They were the representatives which 'society' sent to the committee and, in the absence of any countervailing force, they dominated it. P.A. Tikston and I.B. Glivits of Prodamet and Kutler and Iu. I. Butlerov from the Urals all took an active part in the Metallurgical Committee's proceedings. General Myshlaevskii later confessed that 'I always was of such opinion that the chairman would only be able to work if he was in the closest contact with, and had the full cooperation of, Prodamet, Med' and Krovlia.' He singled out Kutler and Butlerov for particular praise, referring to the former as 'my teacher' and 'the actual director'.[21] The syndicates used the Metallurgical Committee to regulate their interrelations and legitimate their price policies. One of the first measures taken by the Committee was to cancel all unfulfilled orders placed before April 1915.[22] This was much appreciated by the managers of metallurgical enterprises who preferred to fulfil more recent, higher priced orders.

There remains one unanswered question in this affair. Why did the TsVPK propose and defend a project which favoured metallurgical factories and, above all, the syndicates at the expense of consumers' interests? According to Tarnovskii, the central committee was under the influence of 'southern monopolists'. The struggle which Pogrebinskii posited between the central committee and Prodamet was, in Tarnovskii's view, really between the Moscow VPK and Prodamet, with the TsVPK supporting the latter. Tarnovskii acknowledges Moscow's influence in the central committee but asserts that 'the TsVPK not soon and not until the end lost its "southern" face'.[23]

Tarnovskii is basically correct, though the situation was a good deal more complicated than he indicates. Representatives from the Urals – Kutler, Butlerov, P.I. Pal'chinskii, B.N. Pomerantsev and F.A. Ivanov – played as important a role in the central committee as the 'southerners'. Moscow's influence was weak because the leading Muscovites preferred it that way. Guchkov, involved with the formation of workers' groups, did not participate in the debates. Konovalov did, but seems to have deferred to the expertise of the metallurgists. Others on the central committee favoured Prodamet and the Urals syndicates because they appeared to offer a better alternative to a free market or bureaucratic control. Typical of their attitude was Zhukovskii's hymn of praise pronounced to the press. 'Whereas formerly syndicates were regarded with suspicion', he said, 'experience has convinced us that a syndicate is a very good thing.'[24]

Moscow indeed was the centre of opposition to Prodamet primarily because its metalworks enterprises, unlike those of Petrograd, had few financial connections or board members in common with metallurgical factories. Only three days after the TsVPK had sent its memorandum to Polivanov suggesting the creation of a metallurgical committee, Jules Goujon, director of the Moscow Metalworks Company, asked that it be amended. In his own memorandum, tactfully sent in the name of the Society of Factory and Mill Owners of the Moscow Region, Goujon proposed to Polivanov that one-half of the representatives from the TsVPK and the Association of Industry and Trade be drawn from the metalworks industry.[25]

The misgivings of Goujon and other consumers of metal were entirely justified. The metal crisis of 1916–17 and the Metallurgical Committee's attempts to deal with it demonstrated that the syndicate's domination of the state institution was detrimental to the rest of the economy. Contrary to the TsVPK's predictions, supply did not nearly cover the state's insatiable demand for metal. The state consumed six and a half times as much shaped steel in May 1916 as it had in January 1915. By March 1916, the deficit in pig iron was 7 million puds, and in May, Myshlaevskii announced that supplies of pig iron and iron covered only two-thirds of demand.[26] Prices of all metals rose following the renegotiation of contracts with Prodamet's members the previous December. Speculation on the market flourished. Even the Metallurgical Committee's certificates required for the sale of iron became marketable commodities. Its determination of maximum prices for orders to the state was not detrimental to the syndicates in that the prices were based on those already prevailing.[27]

While the TsVPK concentrated on ways to increase coal supplies and workers at metallurgical factories, among the Unions and within the Moscow VPK the demand for price control gained wide currency. The Kadet, A.I. Shingarev, representing the Union of Towns, tabled the motion at a session of the Special Council of Defence in February 1916 and repeated it a month later to the Metallurgical Committee.[28] The Moscow VPK and the Zemgor wanted the regionally based factory councils and not the Metallurgical Committee to do the stabilizing. On 8 May, representatives from several district VPKs met in Moscow and passed a resolution to that effect.[29] The Moscow delegation to a congress of district committees held later in the month was vociferous in attacks levelled against Prodamet, and received the backing of Tereshchenko. Resolutions were adopted for the immediate introduction of a maximum price for iron and a system of inspection organized by the VPKs.[30]

The central committee, seemingly oblivious to provincial dissent, continued to back the syndicates. D.S. Zernov, admitting the seriousness of the metal crisis, contended that while 'public supervision' of Prodamet was impossible, state control was 'out of the question'.[31] Matters came to a head at a session of the central committee's metallurgical section on 25 May. Two Moscow delegates, N.P. Nochevkin and V.N. Pereverzev, reiterated the demands for public supervision of Prodamet and measures to increase output. But the syndicates were represented in

full force. Vvedenskii presided; A.Iu. Rummel', an
executive of Prodamet since its earliest days,
claimed that to hold the syndicate accountable for
shortages was unfair; Tikston warned that 'the
establishment of control over Prodamet would not
change anything because it has been under the
control of the Metallurgical Committee for a long
time'; S.A. Erdeli concluded that 'syndicate' was
an inappropriate term for Prodamet. When Konovalov
spoke in a similar vein, Pereverzev walked out in
disgust.[32]

The TsVPK's support of the syndicates was at the
expense of its own fulfilment of orders. Promised
600,000 puds of metal per month to share with the
Zemgor, it received considerably less throughout
the first half of 1916. Its own materials section,
created in June 1916, distributed orders for over
2 million puds of iron, but by October only
200,000 had been processed of which approximately
half actually reached enterprises.[33] Occasionally, the dolorous effect of relying on the
Metallurgical Committee and Prodamet was acknowledged by the central committee, but it put forward
no alternative.[34] When A.A. Vol'skii, a member
of the committee, advanced an ambitious plan for
the centralization of all syndicates under a Chief
Administration for Metal Supply, Pal'chinskii,
Zernov, Kutler and Konovalov all argued against
it, and the idea was scrapped.[35] When the
Minister of Ways and Communications announced
plans in late 1916 to construct a huge metallurgical plant in Kerch' (on the Crimean peninsula),
the committee's members were prominent in the
opposition. A.A. Bublikov called it 'new industry
without industrialists'. Pal'chinskii advised
relocation to the Far East where private
initiative would not be effected.[36] The plan,
which required the Duma's approval, had not
received it by the February Revolution.

The central committee's experience with non-
ferrous metals was similar to its involvement in
the regulation of iron and steel in a number of
respects. Representatives of the copper syndicate,
Med', were prominent in the purchasing committee
sponsored by the VPK. The processors of copper
opposed this organ just as some metalworks owners
resented the Metallurgical Committee. But the
differences are equally important. Whereas the

committee's Metals Bureau dealt with domestic production and distribution, the purchasing committee directed its attention to foreign supplies of non-ferrous metals. While the TsVPK initiated the process for the supercession of its Metals Bureau, it tried to sustain the foreign purchase committee for as long as it could.

In 1913, domestic mines provided 85 per cent of all copper required by Russian industry, but only 27 per cent of the zinc, and minute proportions of lead, nickel, aluminium and antimony.[37] Despite the importance of some of these metals for war-industrial production, output actually declined after 1914 and imports failed to make up the difference. By August 1915, the Duma's Budget Commission was informed that 'many metals necessary for state defence, such as tool steel, zinc and copper, have either disappeared from the Russian market or are concentrated in the hands of monopolists'.[38]

Two organizations were created to obtain the necessary metals from abroad. One, sponsored by the TsVPK with the apparent backing of the Special Council, was the Committee for the Purchase of Foreign Metals (hereafter Foreign Purchase Committee). Formed in late August 1915, it was run by a bureau consisting of F.A. Ivanov, B.N. Pomerantsev and S.I. Litauer, all of Med', and the Octobrist deputy, I.I. Dmitriukov.[39] The functions of the committee were to aggregate orders from private enterprises, obtain foreign currency, place the orders on the world market, and arrange for the delivery of the metals. The other organization took the form of a special commission for metals under the Ministry of Trade and Industry. Its purpose was 'the organization of the purchase of several metals required by the state'.[40]

The activities of these two bodies were significantly affected by the agreement concluded between Russia's Finance Minister, P.L. Bark, and Britain's Chancellor of the Exchequer, MacKenna, on 17/30 September. According to this agreement, the British government committed itself to setting aside £4.5 million every month for Russian purchases of war materials in the British empire and North America. All orders had to be confirmed by a joint Anglo-Russian Committee in London.[41] While intended to reduce costly competition among the allies, the agreement only heightened it among Russian ministries and agencies such as the Foreign Purchase Committee by placing a limit on Russian orders.

In an attempt to eliminate such competition, the Council of Ministers resolved on 3 November 1915 that

> in the case of foreign market dealings an appropriate organ must be formed directly by the government which does not depend on the cooperation of public-industrial organizations.[42]

The resolution constituted a victory for the Ministry of Trade and Industry which had sought to give its special commission an advantage over the Foreign Purchase Committee. However, the victory was short-lived. On 12 November, Polivanov approved the statutes of the Foreign Purchase Committee which defined its tasks as 'cooperation in receiving metals purchased abroad by factories themselves, and cooperation in receiving foreign currency'.[43] Shortly afterwards, the Ministry of War's Currency Commission upgraded the role of the committee from 'cooperation' to the exclusive right to request up to £1 million sterling each month and distribute it among enterprises. Such defiance of the Ministers' resolution was too much for Prince Shakhovskoi, the Minister of Trade and Industry. 'The Committee has changed', he complained to Polivanov. 'The Committee has become a monopolistic organization for the purchase of foreign metals.' It was necessary, he argued, to transform it into a state institution and revise its membership so that state officials were in the majority.[44] Three days later, Bark added his voice to Shakhovskoi's demand, noting the British government's approval.[45]

The TsVPK's sponsorship of the Foreign Purchase Committee was terminated on 21 January 1916. According to a plan worked out by the Preparatory Commission of the Special Council of Defence, the committee retained all its functions but became a state organ. This did not, however, end the involvement of the TsVPK with the foreign purchase of metal. Two of its representatives remained in the Foreign Purchase Committee and Ivanov, at the request of Polivanov, continued as chairman. Work proceeded as before. By March, the committee had ordered 2.18 million puds of various metals through the Anglo-Russian Committee and provided currency for the purchase of nearly a million puds by private enterprises.[46] The total amount was considered by Ivanov to 'completely cover the requirements...of all factories working for defence', assuming the delivery of the orders in full.[47]

But the consumers of imported metal were not satisfied. Several factory directors petitioned the government for yet another reorganization, one that would include them. They accused the committee of negligence in distributing foreign currency, and implied that Ivanov as a prominent representative of Russia's copper syndicate purposely withheld supplies from abroad in the interests of higher domestic prices. Despite his submission of countermemoranda, Ivanov's position became untenable. In March, General E.K. Hermonius, the chairman of the Anglo-Russian Committee, returned to Russia and threw his support behind the anti-Ivanov group.[48] A month later the *coup de grâce* was delivered. The Foreign Purchase Committee was absorbed by Myshlaevskii's Metallurgical Committee, representatives of processors were included, and Ivanov was relieved of his responsibilities.

Apart from sponsoring the Foreign Purchase Committee, the TsVPK engaged in international commerce to obtain machinery and metal for its own contractors. The financial agreement with Britain and the Anglo-Russian Committee took on special significance in this respect as well. Prior to 30 September, the VPKs ordered tool steel in Britain and sent three representatives, Professor B.A. Bakhmet'ev, S.I. Gavrilov, an engineer, and the Moscow industrialist P.A. Morozov, to the United States to purchase independently or through the Russian Government Supply Committee required materials.[49]

After the agreement, the Supply Committee was reorganized and placed under Major-General Sapozhnikov, who set up headquarters in New York's Flatiron Building. Henceforward, all orders placed with firms in the United States and Canada had to be initially approved by the Supply Committee and confirmed by the Anglo-Russian Committee in

London. The actual negotiations for orders and arrangements for deliveries were handled through the 'good offices' of J.P. Morgan, Britain's - and therefore Russia's - sole agent in North America. The VPKs' representatives, as well as those sent by the Zemgor, played an active and important part in the Russian Supply Committee, constituting the entire membership of its supervisory commission and contributing chairmen to two others. But so circumscribed were the powers of the Supply Committee that the influence exercised by Bakhmet'ev, Gavrilov and Morozov did not give the organization they represented any leverage in tapping the American market.[50]

In search of alternative sources of steel and other equipment, the TsVPK dispatched to Sweden I.I. Liasotskii and, appropriately enough, E.L. Nobel. But most important of all for obtaining foreign credit was representation on the Anglo-Russian Committee. To some within the government, the committees' petition for representation appeared eminently reasonable. For example, Vice-Admiral Rusin, Russia's chief negotiator at an allied conference convened in London in November 1915, wrote that

> The Anglo-Russian Committee should have representatives from the public organizations, to wit, the TsVPK, the Unions of Zemstvos and Towns. These institutions are large purchasers and their interests are not represented by anybody.[51]

Polivanov also championed the public organizations' cause. On three separate occasions, he supported petitions to the Council of Ministers to include their representatives on the Anglo-Russian Committee. Writing to B.V. Stuermer, the new chairman of the Council, Polivanov argued that the committees' presence would be valuable because they had much 'theoretical and practical knowledge of industry's needs' and had already acquitted themselves well in America. On the other hand, their absence would deny them the possibility to place orders abroad, something which he warned 'would be harmful to state defence'.[52] Finally, he threatened to take the matter up personally with the Tsar. Prince Shakhovskoi later took credit for aligning the ministers against the committees and Polivanov. In his view, 'the success of Hermonius as head of the government's delegation to London inspired Guchkov to intrigue against it and demand a "public element"'.[53] The Council of Ministers concluded that the VPKs were guilty of 'an attempt to act not in concert with, but parallel to, governmental authority'.[54] The Tsar showed his sympathies by replacing Polivanov with D.S. Shuvaev as Minister of War.[55]

In the space of a few short months, the Tsarist government, aided by certain entrepreneurial interests and unwittingly by the British government, effectively isolated the VPKs from the administration of foreign orders. After the removal from office of Polivanov, their only real supporter within the higher spheres of government, the committees raised the issue of foreign orders and their right to representation on the Anglo-Russian Committee a number of times, but without success.[56]

6.2 COAL AND OIL

State regulation of coal and oil, the two most important industrial fuels, represented a tortuously slow, step-by-step process of increasing control over their prices and procedures for their distribution. More vigorous measures such as general requisitioning and the introduction of state mines and wells were proposed and discussed in the Special Council for Fuel but never instituted. One of the reasons for this, as well as for the ineffectiveness of price controls, was the strong opposition mounted by mineowners and oil magnates. The VPKs served as a mouthpiece for both.

The central committee's fuel section was the equivalent of its Metals Bureau in terms of its membership and orientation. Formed in June 1915 and consisting of four subsections - coal, oil, wood and peat - it dominated the discussion of fuel within the committee, and, in the absence of opposition from other committees, spoke for the entire organization. The chairman of its presidium was N.F. Fon-Ditmar, owner of a mining equipment factory in Khar'kov and, since 1906, chairman of the Association of South Russian Mineowners. S.A. Erdeli, a major shareholder in the Urals Bogoslovskii Mining and Metallurgical Company and its representative to Med', was the first chairman of the coal subsection. He was eventually replaced by L.G. Rabinovich, an engineer and director of the Grushev Anthracite Company.[57]

N.N. Iznar, the Association of Baku Oil Producers' agent in Petrograd, was the oil subsection's first chairman. After assuming the chairmanship of the military stores section, he was replaced by M.B. Pappe but continued to attend the subsection's sessions. Pappe was chairman of the board of 'Caucasus' which belonged to the Royal Dutch Shell trust. He was one of many representatives from oil companies to participate in the TsVPK's fuel section. M.M. Beliamin, the chairman of the East Asian Oil Company and a director of Nobel Brothers, P.O. Komov and E.L. Nobel himself were from the Nobel Brothers trust. G.A. L'vov was on the board of eleven companies, ten of which were controlled by the Neft' (Russian Society for the Production, Transport, Storage, and Sale of Oil Products, founded in 1883) trust. Finally, the Russian General Oil Corporation, a trust which incorporated the Lianozov concerns and was responsible for the sale of more oil in Russia than any of the other three, was represented by E.A. Vatatsi, A.O. Gukasov and P.O. Gukasov.[58]

The fierce competition among these trusts did not preclude a common front within the TsVPK. Respecting the expertise which the trusts could muster and knowing of no other or no 'higher' form of organization, the central committee deferred to the trusts' representatives on all questions related to the regulation of oil. Consequently, the oil subsection resembled nothing so much as an echo chamber of the Association of Baku Oil Producers.

Russia's coal industry had no syndicate or trust during the war comparable to oil. Never having attained the degree of control which Prodamet and the oil trusts exercised, and seriously weakened by legal complications in 1913-14, Produgol', the Society for the Sale of Mineral Fuel of the Donets Basin, was practically overtaken by the huge demand created by the war.[59] In the event, the Association of South Russian Mineowners acted as a unifying force among major coal entrepreneurs. Its members were well-placed in the VPKs - P.G. Rubin (Ekaterinoslav), A.I. Fenin and G.A. Kol'berg (Khar'kov), and Fon-Ditmar and P.O. Kozakevich (central committee) - and influenced discussions therein on the supply of coal.

Although coal output in the Donets was 181.2 million puds more in 1915 than in 1913, the cessation of imports, the general increase in demand, and the limited capacity of the railways to transport coal resulted in severe shortages.[60] The second half of 1915 was an especially difficult period. Between September and the end of the year, only two-thirds of the coal scheduled to be transported by rail was actually hauled away from the pit heads and depots. Petrograd received 49 per cent of the coal assigned to it in October and Moscow received 40 per cent.[61]

Prior to the creation of the Special Council for Fuel, the regulation of fuel supply was in the hands of S.V. Rukhlov, Minister of Ways and Communications. To supplement the work of existing institutions in correlating supply with demand, Rukhlov in March 1915 established a Committee for the Distribution of Fuel with provincial branches. This committee, which contained representatives from private industry, did not interfere in the production or sale of coal but regulated distribution on the basis of gradations of consumers' importance. The inadequacy of this procedure was apparent to all, but there was no consensus on what to do. Mineowners spoke of the need to increase the number of wagons and workers at the mines. Representatives of the ministries and state railways defended the performance of transport and accused the mineowners of deliberately reducing output.[62] The selective requisition of coal, intended as a punitive measure, was no solution to the problem of coordinating regular loading and train schedules. In fact, it encouraged delinquency among mineowners because the requisition price, although below that on the market, was still above the price according to which the 3 year contracts had been drawn up in 1913.[63]

The Special Council for Fuel, under the Minister of Trade and Industry, addressed itself to a revision of requisitioning practices. On the one hand, representatives of private producers

introduced a plan for the reorganization of administrative organs but no increase in their control
over production or sales. On the other hand, the
ministerial representatives favoured a general
requisition amounting to a virtual state monopoly
over the sale of coal.[64]

Organized private enterprise firmly opposed the
application of a general requisition and the VPKs
were in the forefront of that opposition. The
seventh session of the central committee heard
L.G. Rabinovich, one of its three representatives
in the Special Council for Fuel, summarize and
denounce the idea in no uncertain terms. A state
monopoly, he contended, would mean the violation
of all previous agreements, causing disruption to
consumers. Prices would undoubtedly rise and
quality fall. Given the existence of more than 40
types of coal fuel and over 700 suppliers,
technical complications alone would prohibit an
effective state monopoly.[65] The committee
endorsed Rabinovich's views. In the following
months petitions and delegations opposing the plan
were sent to Shakhovskoi from a number of organizations. Riabushinskii expressed his lack of faith
in the ability of 'bureaucratic elements to
organize this business and expediently carry it
through'. The Moscow VPK, after hearing
Konovalov's report on the central committee's
negative reaction, resolved to inform the Special
Council of its opposition.[66] Metallurgists'
objections stemmed from their fear that the
general requisition of coal would lead to the same
arrangement for metal. On the basis of
Rabinovich's report to their fortieth congress,
the Association of South Russian Mineowners voted
110 to 10 against the plan.[67]

This cumulative pressure militated against the
realization of the state monopoly which the
Special Council for Fuel had approved in principle
in September. For months the council held its
decision in abeyance. It did not go beyond regulating prices for surplus stocks, granting bonuses
to enterprises which over-fulfilled their orders,
and requisitioning stocks of those which failed to
deliver theirs.[68]

But in mid-1916, anxious to take advantage of
the rise in production, ministerial officials
resurrected the idea of a general requisition,

this time in the form of a mixed state-private syndicate known as 'Tsentrougol''. The TsVPK was quick to condemn the new proposal. According to S. Kornfeld', secretary of its fuel section, 'Tsentrougol'' would 'shake the foundations of the mining industry' and cause 'great complications in the long-established relations between consumers and mineowners'.[69] The proposal won praise from Rabinovich but most representatives of the mining industry were concerned about governmental interference and its perpetuation after the war. The Moscow VPK, on the other hand, expressed the fears of consumers who were not represented at all on the syndicate's board. Other VPKs, Khar'kov and Rostov-on-Don, indicated their opposition in a questionnaire distributed by the Association of Industry and Trade.[70] Taking note of the nearly universal unpopularity of the plan, Shakhovskoi buried it in December.[71] Only after the February Revolution, with Konovalov as Minister of Trade and Industry and Pal'chinskii effectively in charge of the Special Council for Fuel, was a state monopoly introduced. Its advisory board contained representatives from state and private railways, the TsVPK, the two Unions, shipping companies, regional workers' soviets, and coal-mine owners.[72]

The VPKs' opposition to the state regulation of coal should not be construed in entirely negative terms. The argument which the central committee consistently advanced against general requisitioning was its irrelevance to the fuel crisis. It was more important, in the committee's estimation, to provide mines with an adequate labour force, to increase the number of trains transporting coal, and to develop new coal fields and alternative sources of fuel.[73] Still, none of these proposals precluded a state monopoly of sales. What did were the objections of industrialists to state entrepreneurship which they regarded as unfair competition. While it was undoubtedly the case that, as the TsVPK's fuel section noted, 'the Special Council for Fuel has only slightly contributed toward the resolution of the fuel crisis', it might have accomplished more had the committee encouraged mineowners to cooperate with it.[74]

As with coal, the TsVPK failed to distinguish

between the requirements of Russia's oil industry and the interests of its oil entrepreneurs. Seeking to provide the conditions for maximum production, it readily accepted the entrepreneurs' arguments that high prices and no state competition would ensure it. Thus, at the committee's eighth session in September 1915, Baron Maidel listed the sale of new land for private exploitation and a price for oil 'favourable to oil suppliers' as prerequisites for an increase in supplies.[75]

The Minister of Trade and Industry, on the other hand, sought to reduce prices and stimulate oil production by the state. In April 1915, Shakhovskoi called together a group of oil magnates, including the two Gukasovs, E.L. Nobel and S.G. Lianozov, to request that they charge no more than 40.5 kopecks per pud for crude oil at Baku. But failing to receive guarantees for adequate transport, materials and intervention in case of workers' strikes, the magnates rejected the appeal for voluntary controls.[76] The price which the state and other consumers paid for this rejection was high. By December oil fetched 51 to 57.5 kopecks in Baku and as much as 1.20 rubles in Moscow.[77]

On 31 December 1915, Shakhovskoi took the unprecedented step of establishing a maximum price for crude oil, 45 kopecks per pud in Baku. This was followed several weeks later by a scale of prices for oil transported to various points throughout the empire. Prices ranged from 53.5 kopecks in Astrakhan' to 96 kopecks in Petrograd.[78] In March, the price of refined oil was frozen, and, finally, on 27-8 April, requisition prices were set at no less than 15 per cent below the maximum prices.[79]

Throughout 1916, oil entrepreneurs besieged the Special Council for Fuel with petitions to raise the maximum prices. They took their case to the VPKs as well. In February, representatives of the oil industry reported to the central committee's bureau that in view of the low price of oil in Baku and high transport costs certain deliveries would be cancelled.[80] Suspicions of shortages were confirmed by the chairman of the Ekaterinoslav VPK. He related that in his city oil had cost 95 kopecks before the imposition of a maximum price but became so scarce thereafter as to fetch 1.65 rubles per pud.[81] The central committe's fuel section drew up an indictment of the price system which was identical with the oil producer's petitions. 'The owners of plots', it said, 'where the rent is excessive and the cost price as much as 45 to 50 kopecks, given a maximum price of 45 kopecks, are compelled to reduce their output of oil.'[82] By November 1916, the producers were giving a most depressing picture of the situation to the TsVPK's oil subsection. A.O. Gukasov, recently returned from Baku, characterized the mood at the oil fields as one of 'extreme despair'. His brother added that 'each day brings new difficulties' and that 'the question of revising the maximum price is urgent'.[83]

In fact, oil production in 1916 was 608.1 million puds compared to 571.4 million the previous year. Railways and barges transported 25 million puds more than in 1915. True, drilling operations declined. But the producers' contention that low prices were the cause hardly explains why in 1915 the rate of decline was greater - 18 per cent compared to 11 per cent in 1916.[84] The producers were thus hard pressed to demonstrate the adverse effect of existing prices. While E.A. Vatatsi lamented that 'we are required to furnish proof of their influence on the reduction of output, a task which is very difficult', P.O. Gukasov admitted that 'of course it would be absurd to assert that today's prices are harmful to everyone. When we say that the prices are harmful we are speaking in terms of the average firm.'[85]

Inflation was the producers' best argument. By January 1917, it was only a question of how much the maximum price would be raised. The ministers of State Control and Ways and Communications claimed an increase of 10 kopecks per pud was sufficient.[86] The oil producers demanded at least 15 kopecks, and on 20 January a majority in the Council of Ministers agreed. The new price of 60 kopecks plus corresponding increases at points of distribution went into effect on 18 February.[87]

Concerning proposals for the state's production of oil and its monopolization of sales, the TsVPK exhibited an aversion worthy of the oil producers'

association. Approval for state monopolization of
production and distribution was voiced by a single
delegate, from Samara, to the VPKs' second
congress. He considered this the only way 'to
break the monopoly of Nobel, Shell, and Oil' (the
Russian General Oil Corporation).[88] But this was
not a popular demand. Three times during the
congress he was deprived of the floor when he
tried to speak on oil. Konovalov, the chairman on
two of those occasions, had spoken his mind on
state monopolies two weeks earlier. Addressing the
Duma on 19 February 1916, he ridiculed the tendency of the government to engage in 'petty control
over the economic activity of the individual'. He
disputed the claim that state monopolies saved the
taxpayer money and concluded with a paean to
'private initiative, private energy, and private
strength [which] must be given complete freedom'.[89]

Shakhovskoi's project for state oil production
sought to reduce the power of private monopolies.
Introduced into the Special Council for Fuel in
June 1916, it called for a total expenditure of
53.7 million rubles over a 6-year period to
develop sites capable of producing up to 50
million puds per year. The initial selling price
was to be 36 kopecks per pud.[90] On the same day
that the project was announced, M.B. Pappe and
P.O. Gukasov complained to the TsVPK's bureau that
it would undermine private industry.[91] Konovalov's noncommital response convinced the Zemgor
organization that 'the oil kings' efforts to attract the central committee are a failure'.[92]
But the Zemgor had not taken account of Konovalov's long-standing commitment to 'free enterprise'. Three years earlier, during the Duma's
consideration of an almost identical project which
arose in connection with the pre-war 'oil famine',
Konovalov had stated:

in my opinion, this project of the government is
absolutely bankrupt both in the principled as
well as in the practical sense....I personally,
gentlemen, am a firm opponent of the treasury's
expansion of economic operations, a policy...
which is at cross purposes with the interests of
the economy as I understand them....But aside
from the financial-economic motive, it is

impossible not to recognize that any increase in the economic might of our government entails an increase in its political power...and this cannot in any way be considered desirable.[93]

His remarks to the Duma in February 1916 suggest that regardless of the war his opinion had not changed. In any case, the TsVPK's *Izvestiia* concluded that Shakhovskoi's project

> creates a threat to the entire Russian oil industry and endangers the future of Baku. The appearance of the treasury in connection with its exclusive right to select the best land and its exemption from the burden of taxes will result in the curtailment of the influx of capital and the development of private initiative.[94]

The article advocated the formation of a special company to which state land would be granted without auction. This proposal had been raised earlier by I.V. Titov, a Progressist deputy in the Duma, and before that by P.O. Gukasov and S.G. Lianozov.[95] After repeatedly opposing it, the government yielded and accepted the plan in February 1917. Its implementation was interrupted by the February Revolution.

6.3   COTTON

If the role of the War-Industries Committees in the regulation of metals and fuel was supportive of those industrialists whose enterprises produced them, then the Moscow committee's cotton section was the protagonist in the struggle against bureaucratic control. The extractive and metallurgical syndicates and trusts sent their representatives to the corresponding sections of the TsVPK, but the Moscow committee's cotton section was a syndicate, albeit in a disguised form.

The main concerns of Russia's cotton manufacturers after the initial mobilization were the supply of raw cotton and its price. Imported cotton, which constituted almost half of Russia's pre-war supplies, dropped in quantity from 13.5

million puds to 7.3 million in the first year of
the war. The loss was compensated to some extent
by a bountiful Central Asian harvest which
provided 17.3 million puds in 1914/15 compared to
14.2 million the previous year.[96] Supplies, in
fact, exceeded the average for the preceding 5
year period. But demand was greater, and this
worked to the advantage of suppliers. Without the
imposition of any controls, prices rose from 16
rubles per pud in June 1914, to 24 rubles in
February 1915, 28 rubles in April, and 32 rubles
in June 1915.[97]

The necessity for halting this trend induced the
Society of Cotton Manufacturers to petition the
Minister of Trade and Industry to intervene.[98]
Obtaining the necessary powers from the Council of
Ministers, Shakhovskoi duly established on 7 July
1915 a Special Committee for the Supply of
Factories with Raw Cotton.[99]

The Committee on Cotton, like many organs
created at this time, was of a mixed state-private
character. Cotton manufacturers, suppliers of raw
cotton and government officials comprised the
committee, which held its sessions in Moscow. The
manufacturers were the most powerful group not by
virtue of numbers but because they occupied the
most influential positions. In N.I. Guchkov,
Aleksandr's brother, they had a highly sympathetic
chairman. N.D. Morozov, board member of the
Bogorodsk-Glukhov Mills, was vice-chairman. St. P.
Riabushinskii and V.P. Rogozhin, the director of
the Tver Mills, chaired several of the committee's
ad hoc commissions. They were also respectively
chairman and vice-chairman of the Moscow VPK's
cotton section, though not until 1917 did they
officially represent it in the Committee on
Cotton.

At its first session on 13 August 1915, the
committee set the price of raw cotton not at the
going rate of 32 rubles per pud but at 8 rubles
less.[100] This, coupled with the fact that prices
for cotton yarn and fabrics were not to be regulated for another 6 months, goes a long way
towards explaining the manufacturers' sizeable
profit rates during 1915. True, evasion of the
price limitations for raw cotton was widespread.
According to Morozov, violations reached such
proportions 'that if energetic measures are not

taken against them immediately, it is possible that the whole business of the committee will be undone'.[101] But manufacturers absorbed the circumvention of the regulatory apparatus by raising their own prices. Even before the establishment of a maximum price for raw cotton, it was asserted by M.V. Chelnokov, the mayor of Moscow, that 'there has never been a time in the history of Russia when the manufacturers were doing so well as at the present moment'.[102] Profits for 91 cotton mills rose from 55.2 million rubles in 1914 to 134 million in 1915, or in terms of returns on fixed capital, from 15.8 per cent to 37.9 per cent. They paid an average dividend of 11.4 per cent in 1915 compared with 5.8 per cent in the previous year.[103] Even taking into account the lower purchasing value of the ruble at home and abroad, one would have to agree that 'the profits of the textile industry surpassed the normal level of income during peacetime'.[104]

The Moscow VPK's cotton section worked in tandem with the Committee on Cotton to ensure manufacturers high profits. While the latter sought to reduce the price of raw cotton, the former, as expediter of the GIU's orders, sanctioned the prices charged by manufacturers.[105] This convenient arrangement was threatened by the decree of 17 December 1915 which required the Committee on Cotton to establish maximum prices for yarn and fabric.[106] However, the committee was in no hurry to comply with the decree. On 13 January, it set up a special commission to determine 'the *possibility* of setting prices for yarn and fabric', appointing among others N.I. Guchkov, Morozov and Rogozhin.[107] In the meantime, the cotton section continued to distribute orders as before. Only in late February was it informed by the GIU that henceforth the Committee on Cotton would establish prices for its orders subject to review by the Minister of Trade and Industry.[108]

Lest this cause concern among manufacturers, Rogozhin had talks with N.I. Guchkov, and related to the cotton section their outcome:

N.I. Guchkov, having already contacted the GIU, explained that the nature of relations...will remain as it was. In those cases when the cotton section requires the sanction of the government

for giving its decisions based on compulsory
power, the committee on cotton will readily give
such sanction. The cotton section only has to
inform the committee of its work. As far as
prices are concerned, the cotton section is
required as before to elucidate them with mill
owners but now must present them for confirmation to the Committee on Cotton. In view of the
fact that a number of members of the committee
are foreign to industry and lack competence in
questions of mill production, N.I. Guchkov
agrees to form a commission to confirm the
prices for Quartermaster orders. Only the most
competent people will take part.[109]

The commission contained seven members including
Rogozhin, St. P. Riabushinskii, and two other
manufacturers.

Only on 2 April and 13 May 1916 were maximum
prices announced respectively for yarn and certain
categories of fabric sold to the army. Most
articles sold on the market remained outside any
limitations. Thus, as the Moscow VPK admitted in a
memorandum, 'the mill-owners compensate themselves
for selling their yarn to the Army Supply Department at a lower price by selling their produce at
a higher rate in the open market'.[110] Even after
the price of yarn was frozen for most market
transactions in July, the dichotomy between market
and military prices persisted. The market was
described by one journal as 'completely disorgan-
ized'. 'Prices are so arbitrary', it noted,

that naming a standard is impossible. Pure
fabric enterprises without spinning mills cannot
calculate for the morrow and are forced to pay
for non-regulated varieties of yarn from 100
rubles per pud.[111]

Of course, weavers were as much to blame for this
bacchanalia of speculation; they were willing to
pay such prices so long as they could dump the
finished product on the market at a price of their
choice.[112] Thus, consumers in the rear and
soldiers at the front were the only losers, the
former paying dearly for clothing and the latter
standing not to receive adequate supplies of
blankets, bandages, overcoats and underclothes.

The role of the Moscow VPK's cotton section in this sordid business is ambiguous. On the one hand, it tried to minimize the damage done to the state by the dichotomy of prices. It distributed 2.2 million puds of surplus raw cotton in March 1916 in direct proportion to the percentage of state orders previously fulfilled by individual enterprises.[113] It drew attention to the relatively small contribution of Petrograd's textile enterprises to the fulfilment of state orders, recommending a more equitable distribution of the burden.[114] In terms of its own area of competence, the enterprises of the central industrial region, it arranged for the distribution of state orders on the basis of an equal share to every enterprise.[115]

On the other hand, a case has been made that the section deceived the Committee on Cotton and later historians as to the extent to which cotton manufacturing enterprises worked for the army.[116] The opportunity for exaggeration presented itself in February 1916 when the section submitted figures to a commission of the Committee on Cotton in preparation for the section's distribution of surplus cotton. According to Morozov, 'the commission did not have the opportunity to check the information presented by the cotton section of the VPK and it cannot take any responsibility for its verisimilitude'.[117] The motives for deception were obvious. The greater the proportion of fabric sold on the market the greater the flexibility of prices and opportunities for profit; yet, only if an enterprise accepted above a certain percentage of orders for the army could it receive privileges in obtaining raw materials and fuel. On 11 January 1916, the cotton section announced that the minimum allowable load was 40 per cent. Was it a coincidence, then, that in its report to the TsVPK in February the Ivanovo-Voznesensk VPK gave as 40 per cent the city's textile output for the army?[118] An additional motive was revealed by Iu. I. Poplavskii. Addressing his colleagues in the Moscow VPK, he warned that if sufficient orders were not forthcoming from the GIU, 'we can expect workers and employees to transfer to those enterprises which can extend them deferments'.[119]

Difficulty arises in amassing evidence that

would demonstrate deception on the part of the
cotton section. According to St.P. Riabushinskii's
report of October 1916, two-thirds of the fabric
output of the central industrial region was
devoted towards the army's needs.[120] When it is
recalled that this period saw the highest
proportion of goods going to the army, and that
that proportion was, as far as textiles were
concerned, higher in the central industrial region
than in Petrograd, this figure does not seem
implausible.[121] As far as Ivanovo-Voznesensk is
concerned, the VPK's figure which refers to
production in February 1916 does not preclude a
much lower rate, even as low as 13-15 per cent,
for almost 2 years of war up to June 1916.[122]

The evidence for claiming that the cotton
section deliberately deceived the state is at best
suggestive. But such was the nature of the VPKs'
brand of regulation and control that price dispar-
ities between goods destined for the army and
those for the market remained, with a result that
by September 1916, 35 of 64 spinning mills had
failed to turn over to weaving factories the
amount of yarn designated for the army by that
month.[123] When in January 1917 the cotton
section was stripped of its discretionary powers,
it was of immediate concern to the Society of
Cotton Manufacturers.[124] The Society addressed a
memorandum to Prince Shakhovskoi asking for a
reduction in the level of production for the army
but an expansion of privileges for enterprises
working entirely on defence orders.[125] The
memorandum may be interpreted as an indication
that the Society's members had not been subjected
to the strict observance of existing regulations
and that they expected greater vigilance from the
Ministry of Trade and Industry.

6.4 EVACUATION AND RAIL TRANSPORT

The exclusion of the VPKs from the Special Council
for Transport limited their participation in the
regulation of rail freight to local organs and
special commissions. Members of several VPKs
attended sessions of the Special Council but
either as observers or as plenipotentiaries of the
Special Council's regional committees for the

regulation of freight transport.[126] In addition, the committees took part in evacuation and relocation commissions. As the evacuation of enterprises from areas near the front was the first problem confronted by the VPKs in relation to transport, it will be considered first.

Whether because of its unwarranted confidence in the fighting capacity of its troops or through sheer oversight, the Russian General Staff had made no provision for evacuating industrial enterprises threatened with destruction or capture.[127] In some cases, it was the VPKs which took the initiative. The central committee alerted the General Staff of the necessity to evacuate Belostok (see p. 79). On 25 July 1915, the VPK of the Northwest Region, with headquarters in Vilna, established an evacuation commission. The commission petitioned the regional military commander to provide 100 freight cars per day until evacuation of the city could be completed. Trains began to arrive on 11 August, and within 2 months 166 enterprises had been evacuated.[128]

As far as the rest of Russia was concerned, the evacuation of enterprises and people from the western border areas was only the beginning of the problem. Refugees poured into the cities of the interior compounding already serious food and health problems.[129] The situation in Moscow was particularly grievous. It was to Moscow, the heart of the rail network, that a large proportion of evacuated freight found its way.[130]

> Aside from the fact that the wagons containing this freight are completely immobilized, the Moscow stations are so overcrowded with them that they are being sent to other lines. A situation has been created where all this freight, among which are multitudes of machines badly needed for defence work, is preventing the proper circulation of trains....Valuable machines and machine-tools which require the most careful attention are left on the naked ground under the open sky.[131]

This, the Moscow VPK's assessment, is borne out by available statistics. While 359 wagons per day brought foodstuffs to Moscow in April 1915, in August 307 arrived on the average, and in September only 126. Coal supplies from the Donets Basin arrived in 374 wagons per day in July but a mere 214 in September.[132]

Moscow attempted to deal with this problem outside the official channels. On 28 August, the VPK set up a special commission under I.I. Olovianishnikov to register, sort and find storage facilities for evacuated freight. By the middle of November, the commission had registered 1139 wagons at stations in and around the Moscow network.[133] The ultimate aim of the commission was to locate the owners of the freight and facilitate the reconstruction of their enterprises. Pursuant of this aim, it relayed petitions for loans and subsidies and supported the right of Jewish entrepreneurs to live in Moscow.[134]

To ease the pressure on Moscow and at the same time foster industrial development in their own areas, other VPKs became involved in relocation work. M.G. Braikevich, chairman of the Odessa VPK, graciously offered to relocate in his city any textile enterprise evacuated from Belostok. The Simbirsk committee advertised the industrial opportunities of the city in terms of its mid-Volga location, its rail connections, the low prices and availability of food, and the climate.[135]

The government was clearly overwhelmed by the magnitude of evacuation and therefore only haltingly intervened. In September 1915, the Special Council of Defence added to its already burgeoning bureaucracy an evacuation commission, placing Rodzianko in charge. Rodzianko's commission classified those of the Moscow and other VPKs as organs 'preparatory' to the formation of official subcommissions.[136] Nothing, however, was mentioned about who was to determine the extent of 'preparatory' work or when it would terminate.

Thus, in mid-November when the chairman of the Moscow factory council informed the VPK that an official evacuation commission had been formed, the committee refused to acknowledge its authority or disband its own commission.[137] It continued to collect information on wagons and notify owners of their whereabouts until March 1916 when it was liquidated. In general, the undoing of the VPKs' commissions was due to a combination of official concern about institutional parallelism, and the fact that far fewer enterprises required evacuation in 1916 than during the retreat of the previous summer and autumn.

The failure of the committees to secure representation on the Special Council for Transport did not prevent them from advising the state on requirements for rolling stock production and proposing improvements in the administration of the railways. Intended as much to impress upon the state the irrationality of its procurement policy as to underline the colossal significance of the railways, the central committee called for the production of 1700-2000 locomotives and 60,000 wagons in 1916. This was, in fact, three times as many locomotives and twice as many wagons as were actually produced. Part of the reason for the disparity was the concentration on more profitable munitions production by large engineering firms.[138]

As orders placed abroad for rolling stock and rails were slow in being fulfilled, what was available was strained beyond capacity. In the latter half of 1915, 10 per cent more freight was carried than during the same period of 1914, although the number of locomotives and wagons in use by the end of the year remained almost the same. In 1916 the railways carried 2.6 billion puds more than in 1915 but the amount of rolling stock in operation by the end of the year was 20 per cent less than at the beginning.[139] The front had extracted its required materials from the rear but at a price which the rear literally could not pay.

Compounding the stresses and strains experienced by the transport system was the inefficiency of the state's regulatory apparatus. The fact that actual deliveries rarely conformed to schedule was perhaps to be expected under the circumstances. But the prevailing arbitrariness provided a golden opportunity for bribery which, apparently, few railway officials could resist. Other than alerting the public to this situation, the VPKs attempted to involve it - or rather its self-styled representatives - in the administration of rail transport. The idea of 'public supervision' (*obshchestvennyi kontrol'*) of the rail network was first proferred by Riabushinskii in November 1915.[140] It was taken up by the Moscow VPK's

transport section with special vigour after the
new Minister of Ways and Communications, A.F.
Trepov, excluded all members of public organizations from the Provisional Regulatory Committee
on Railways, yet another state organ which sought
to make order out of chaos.[141] The VPK's scheme
won the approval of several local officials
including V.D. Voskresenskii, chairman of the
Moscow regional transport committee. According to
a plan which he tabled in January 1916, members of
exchange committees, consumer and cooperative
societies, as well as the Unions and the VPKs
would twice a month submit reports to the regional
transport committees on the arrivals, unloading
and departures of all freight trains. These
reports were to be based on unscheduled 'flying'
inspections of stations and the books of railway
officials.[142] The regional committees would then
discuss their findings with the railway
administration.

Approved in principle by the second congress,
the plan was turned over to the TsVPK's transport
section for elaboration. Bublikov, in introducing
it, expressed mild approval. Another speaker was
less charitable, referring to 'many obscurities
and omissions' in the plan. F.I. Shmidt, chairman
of the board of the Riazan'-Urals Railway, a
private line, was understandably hostile. 'Too
large a number of supervisors, often completely
ignorant of the business and procedures of
stations but none the less anxious to interfere in
them, this is also in its own way an evil', he
averred.[143] The section decided to appoint a
special commission to reconcile the opposing
sides. But in May 1916 the commission reported
that it was unable to arrive at a unanimous
decision. Four representatives of private railways
steadfastly opposed wider public supervision. The
section resolved to forward the plan without its
own endorsement to the Special Council for Transport.[144] The minutes of the latter organization
contain no mention of it.

6.5 MANPOWER

Russia's most precious resource but the one most
easily squandered during the war was its manpower.

Underlying the shortages of materials and fuel was
a lack of qualified workers. This was a condition
which existed in pre-war years, but the war intensified it. The army grew to an unprecedented size
- 6.4 million men in 1914, 11.7 million in 1915,
and 14.4 million in 1916. The latter figure
represented 36 per cent of the total male population of working age. Ten million peasants were
mobilized, but especially in the early mobiliz-
ations factory workers, whether skilled or
unskilled, whether employed in essential or
nonessential industries, were not spared.[145]

During the first year of war, as it became
obvious that each worker who entered military
service represented an increase in the demand for
supplies and a corresponding decrease in industry's ability to provide them, there was an
expansion of the system of deferments. In November
1914 the Minister of War was not willing to go
beyond a consideration of deferments for skilled
workers employed in enterprises which accepted
orders for military equipment. By the following
June, however, the General Staff included all
factories supplying these enterprises with raw
materials and semi-finished articles.[146]

Still, industrialists were not satisfied. The
call to mobilize industry, so far as it concerned
the question of manpower, meant a reversal of
priorities. Not only were wider sections of the
working population to be exempted from military
service, but all skilled workers were to be
returned from the front to their factories. The
latter demand was voiced repeatedly by Jules
Goujon. Having spent some time at the front,
Goujon wrote to the Ministry of Trade and Industry
that 'everywhere, I heard the request: "Take back
your workers but give us guns and above all
shells."' To the VPKs' first congress he moaned,
'each day we are sent new orders but all I can do
for the past month and a half is send them back
because I don't have any workers'.[147]

Goujon found much sympathy among the delegates.
They carried a resolution demanding the return of
all skilled workers, arranged for a personal
delegation to Headquarters, and directed the TsVPK
to raise the matter in the Special Council. But at
least until July 1916, the committees were,
according to P.P. Kozakevich, 'singularly unsuc-

cessful' in obtaining the return of technicians or skilled workers.[148] Between September 1915 and May 1916, factory owners requested through the TsVPK's labour section the return of 6626 'specialists'; the section passed on 1507 of these to the military authorities, and only 761 were honoured.[149] Apparently, the General Staff did not agree with the second congress's assertion that 'every skilled worker is incomparably more useful for defence in his specialty than at the front'.[150]

The deferment of worker-reservists involved the committees in a more resolute if only slightly more successful campaign. Beginning in August 1915, the central committee served as an intermediary between factory owners and the military, passing on lists of proposed deferments from one to the other. This was a burden with no rewards. The TsVPK had no discretionary power and, in the view of one Petrograd industrialist, 'only complicated matters'.[151] The industrialist advocated a return to the previous system which depended on factory inspectors, but the TsVPK had other ideas. Its labour section, in preparation for the annual call-up of recruits, proposed a sweeping reform of existing procedures. Under the banner of 'decentralization', it recommended the regulation of deferments by the public organizations. Deferments were to be granted only to workers, and technical and administrative personnel employed at factories 'working for defence'. This designation was to be selectively applied by special *uezd* committees and confirmed by provincial boards. It was these committees, liberally sprinkled with representatives from district and local VPKs, which would exercise discretion over factory owners' lists. Any major decision regarding recruitment policy was to be determined by a central board nominally under the General Staff but containing 12 representatives of the public organizations as opposed to 11 governmental and military officials.[152]

The government accepted the principle of decentralization but interpreted it quite differently. It adopted the plan for a central board with provincial branches but reduced the number of representatives from the public organizations to a minimum. Compared to 17 civil or military officials on the central board for deferments, the public organizations were given 4 places. Each of the 592 designated *uezd* committees reserved 1 seat for the corresponding district or local VPK instead of 3 as recommended in the original proposal. Finally, more objective criteria for determining enterprises 'working for defence' were introduced which narrowed the discretionary powers of the *uezd* committees.[153]

In addition to their efforts to return skilled workers from the front and defer those not yet called up, the VPKs looked to alternative sources of labour power. The first congress was remarkable in this respect, for it was then that the shortage of workers was most keenly felt. Kozakevich and Fon-Ditmar wanted the Association of South Russian Mineowners rather than the Ministry of Trade and Industry's factory inspectors to distribute prisoners of war to the area's mines and factories. Another delegate considered the VPKs eminently suited for this task. The general climate of opinion, as expressed in the congress's resolutions, favoured the abolition of restrictions on night-work for women and children, the employment of children under 15 years of age, and the recruitment of Chinese workers by mines and factories west of the Volga.[154]

The government eventually fulfilled these desiderata. A decree of 19 October 1915 revoked a number of restrictions concerning overtime work for women and children. According to a resolution of the Council of Ministers passed in June 1916, Chinese and Korean workers could be employed in the Donets Basin.[155] Though factory inspectors retained control over prisoner of war labour, the number of prisoners at factories and mines in South Russia soon exceeded the legal limit of 15 per cent. In coal mines they constituted 15 per cent of the total labour force in May 1916 but 25 per cent as of January 1917; in the iron ore mines of the Krivoi Rog, they comprised 42 per cent already in October 1915 and a year later, 55 per cent; in the metallurgical factories of South Russia and the mines of the Urals, the prescribed limit of 15 per cent was exceeded in April 1916 and the number continued to rise thereafter.[156]

Ultimately, neither a piecemeal system of deferments nor the influx of poorly motivated and unskilled workers could compensate for the hundreds of thousands sent to the front. Industrialists, not without reason, regarded both the intensification and extension of the work process as solutions. *Utro Rossii* recommended the wider application of piece-work.[157] The Moscow VPK published a report advocating the three-shift system which when applied to oil drilling in Grozna had boosted productivity by 25 per cent.[158] At the VPKs' first congress, one over-zealous coal-mine owner demanded a 7 day working week and a reduction in the number of holidays.[159] More ominously, several delegates put forward proposals for the militarization of workers.

Militarization, the ultimate weapon at the disposal of a state to control its labour force, was an extremely controversial issue during the First World War. No state imposed complete militarization, though Germany with its Auxiliary Service Law of December 1916 and its concept of total warfare arguably came closest.[160] As far as Russia was concerned, it has been asserted that 'even without the legislative introduction of the militarization of labour, the Tsarist government introduced militarization in practice'.[161] This is true but only in a limited sense. From the outbreak of the war, numerous trade unions were disbanded, several social democratic press organs were shut down and many party activists were arrested. Restrictions were imposed on labour mobility and in the course of 1915 and 1916 they were extended. But precisely because no comprehensive programme for militarization was ever implemented, the whole amounted to less than the sum of its parts.

That no system of militarization was introduced in Russia may be attributed to a number of factors. Rivalry between the War Ministry and the Minister of Trade and Industry over the responsibility of military officials and factory inspectors was never resolved and probably contributed to the fact that neither's proposal for militarization was approved by the Council of Ministers.[162] Opposition within the Duma to militarization surfaced as early as July 1915 and persisted thereafter. B.V. Stuermer, Minister of Internal Affairs from March to July 1916 and chairman of the Council of Ministers for much of that year, was convinced that the Duma would reject any legislative proposals for militarization. Other ministers feared that the invocation of Article 87 in this case would exacerbate already strained relations.[163] Although strict labour discipline appealed to many industrialists, the state's intervention in hiring and firing, wage rates, and the requisitioning of men and materials from enterprises - all of which militarization entailed - did not. Finally, the government, the Duma and industrialists could not overlook the possibility that militarization would touch off rebellion in the factories. In more general terms, neither workers nor industrialists were sufficiently integrated into the state's administrative apparatus to permit the kind of politicking between representatives of the bureaucracy, industry and labour that appears to have been a precondition for the implementation of militarization in other countries. It is significant in this context that the group of industrialists which enjoyed the closest relations with the bureaucracy, the Petrograd Society of Factory and Mill Owners, advocated the militarization of workers and remained largely outside the VPKs. The alternative solution to the labour question which the VPKs devised deserves a separate chapter.

6.6 CONCLUSIONS

The measures at the disposal of all wartime governments for regulating their respective economies were numerous. They ranged from relatively mild ones such as the establishment of maximum prices and partial requisitioning to the expansion of state-owned enterprises, state monopolies for the marketing and production of goods and the militarization of workers. It is remarkable that in Russia, where on the eve of the war the state owned and controlled a larger share of the nation's economic resources than any other state, regulatory measures were adopted during the war in the least systematic fashion and to a lesser extent than, for example, in Britain, Germany and France.

Part of the reason for this was the staunch

opposition to state regulation by Russian industrialists and organizations such as the VPKs.
Certainly Russian industrialists did not exercise
a stronger influence on state policy than their
British, German or French counterparts. Quite the
contrary was true. The point is that in other
countries that influence was exercised by industrialists within or in conjunction with their
governments. In Russia, industrialists both within
and outside the state's regulatory organs opposed
state control while the state's officials refused
to cooperate with industrialists' often politically suspect initiatives.

The question was not whether industry and
commerce should be regulated, but according to
which principles and by whom. The VPKs identified
themselves with the interests of the country but
tied them to the interests of private enterprise.
Although this was a reversal of the priorities of
their parent organization, the Association of
Industry and Trade, the equation remained the
same.

At first, the central committee argued *prima
facie* that 'without the independent initiative of
industrialists themselves the successful conduct
of mobilization is impossible'. This meant that
compulsory measures would only be permissible 'for
separate cases when an enterprise neglected its
social responsibility'.[164] When it became apparent that independent initiative and social
responsibility could not resolve the imbalances,
shortages and contradictions generated by the war,
the VPKs advanced the notion of public regulation,
that is, regulation by the public organizations.

The nature of public regulation, of course,
depended on which groups were representing themselves as the public. As industrialists and
particularly representatives of syndicates were
prominent in the corresponding sections of the
TsVPK, public regulation turned out to be a thinly
disguised form of self-regulation.

The committees initiated campaigns for public
supervision and decentralization, proposed wider
'public' representation in organs already formed
by the state, and sponsored several of their own
organs. They generally opposed price controls and
if prices were regulated, they worked for their
upward revision. Partial requisitions were

acceptable, especially if carried out with the assistance of the committees or their representatives. General requisitions and state monopolies were considered repressive, costly and unnecessary. The same attitude prevailed with respect to the militarization of workers.

The committees' form of public regulation was based on the assumption that industrialists could harmonize their interests and administer the economy more efficiently than the government. But the very activity of the VPKs reveals the weakness of this point. Prodamet, backed by the TsVPK, regulated the supply of iron and steel in the interests of its constituent enterprises which, in the view of the Moscow and other provincial VPKs, was not in the interests of consumers. Controlling the distribution of currency for foreign purchases of metal, the TsVPK incurred the enmity of those firms deprived of independent purchasing rights. Railway companies showed little sympathy for the call for public supervision of transport.

The conflicts between the government and the committees, within the VPKs, and between them and groups of industrialists sabotaged the effective regulation of the war economy which in principle all agreed was necessary. Far from fulfilling their ambition to 'introduce the organizational idea into the chaos of Russian economic reality', the VPKs contributed to the prevailing chaos.[165]

7 The Labour Question – the Workers' Groups of the War-Industries Committees

The workers' groups* – examples of responsible social democracy or 'bourgeois *Zubatovshchina*'; instruments of social peace or organizational centre for revolution. These epithets and others equally contradictory have been used to describe the organs of worker representation sponsored by the VPKs. No other aspect of the committees has attracted so much attention from historians but remained as elusive as these groups of workers' representatives.

The formation of the workers' groups and their role in the politics of industrial mobilization must be seen in the context of long-standing debates within both the social democratic and bourgeois oppositional movements over which would exercise 'hegemony' in the struggle against the Tsarist government. As early as 1898 it had been observed that 'the further to the east one goes in Europe the weaker in politics, the more cowardly and the meaner becomes the bourgeoisie, the greater are the cultural and political tasks that fall to the lot of the proletariat'. Yet how this dictum, composed by P.B. Struve and incorporated into the manifesto of the Russian Social Democratic Workers' Party's first congress, was to be applied – whether it was ever possible to enter into an alliance with or support the 'weak' and

* The Russian *rabochie gruppy* has been rendered as workers' groups rather than 'labour groups' (preferred by some historians) to distinguish them from the committees' labour sections (*rabochie otdely*) which provided information on the supply of and demand for factory workers, and to indicate that the groups consisted entirely of workers or worker-intelligentsia.

'cowardly' bourgeoisie and if so under what conditions – remained a matter of dispute for years to come. It figured prominently in the split between Bolsheviks and Mensheviks, in the two factions' positions during the 1905 revolution, and in the lessons which each drew from the failure of that revolution.

Similarly, the question of the efficacy of an alliance with the left arose within the nascent bourgeois liberal camp, organized into political parties after 1905. Recognizing the dangers inherent in such an alliance and chastened by the violence of 1905, the liberals none the less were aware that without the expansion of political freedom the militancy of the working class could be revived with even more serious consequences.

The decision by the War-Industries Committees' first congress to invite workers' representatives revived and significantly sharpened these respective debates. At once, the question of 'whether' or 'under what circumstances' took on the more concrete aspect of how to respond to an offer of collaboration. With the election of workers' groups in Moscow, Petrograd and other major cities, collaboration had become a *fait accompli*. The workers' groups comprised the only national network of labour activists in Russia linked with a predominantly bourgeois organization, and were also, with the partial exception of the Workers' Insurance Council, the sole legal above-ground organization dealing with the day-to-day problems of the industrial working class during the war. As such, their existence and their activities remained a contentious issue for both the liberal and social democratic movements until the February Revolution.

Despite the importance of this issue and the light it can shed on the alignment of political forces both before and after the overthrow of the monarchy, no western historian has systematically analysed the workers' groups or their relationship with the War-Industries Committees. For the most part, they have been seen as adjuncts of the committees and a compatibility of interests has been assumed.[1] Soviet historians, while giving the workers' groups more prominence, have interpreted them as an example, perhaps the prime example, of 'class treason' committed by the

Mensheviks during the war.[2] But this charge, based on the notion that the groups 'tried to distract the proletariat from the revolutionary struggle', is merely the negative expression of the prevailing Western view. What is therefore required is a consideration of the motives behind the invitation extended by the VPKs' leaders and the aims of those workers' representatives who accepted it, the actual involvement of the workers' groups in the regulation of industrial relations, and their reaction to the surge of militance and political activity among factory workers.

7.1 CLASS COLLABORATION VS 'BOYCOTTISM'

In the months preceding the formation of the VPKs, Russia experienced the most serious outbreak of labour disturbances since the beginning of the war. This unrest was widespread, had many causes, and took a variety of forms. From late March through April 1915, thousands of mine and metal workers in the Donets Basin took part in generally short-lived strikes. Demands included wage increases, a lessening of fines, better conditions in company dormitories and even the ousting of managers of German extraction.[3] The height of the Russian army's retreat from Galicia and Poland coincided with major disturbances in Moscow directed for the most part against suspected German-owned property, a general strike among textile workers in Ivanovo-Voznesensk, and a clash with troops in nearby Kostroma which left several workers dead and scores wounded. There were as yet few strikes which factory inspectors could characterize as political, but the fact that an unprecedented number of workers were disrupting the production of goods deemed essential to the war effort was intolerable to many industrialists and state officials.

By the time the VPKs convened their first congress in July 1915, the militarization of workers, or, more euphemistically, of industry, was being advocated by both bureaucrats and industrialists. During the congress' debate on this question, delegates from South Russia and Petrograd, citing the high labour turnover and the

fall in productivity experienced by their industries, recommended militarization as the only
solution. But they were in a minority. The
alternative put forward by several provincial
delegates and members of the Imperial Technical
Society was to give workers a stake in the war by
providing mechanisms for the articulation of their
grievances. They spoke of the necessity for taking
conciliatory steps such as campaigning for the
reopening of trade unions and the release of the
Bolshevik Duma deputies from Siberian exile. For
as one provincial delegate remarked, 'a class
which does not have the possibility to organize,
which does not have any leaders, must remain dark
and humble'.[4] It was to provide such a possibility that the congress resolved to invite ten
workers' representatives to join the TsVPK and to
open up other committees to workers.

The initial problem confronting the committees'
leaders in the implementation of the resolution
was who would represent the workers. Previous
attempts to link workers with the public organizations had done little to remove the suspicion
with which each viewed the other. The most recent
effort had been a conference on the cost of living
sponsored by the Union of Towns in mid-July. On
the one hand, M.V. Chelnokov, chairman of the
Union, had opposed the invitation extended to
representatives of cooperatives, sick benefit
funds and the few extant trade unions. On the
other hand, these representatives of labour were
openly contemptuous of the Union's assumption that
a solution to the crisis of inflation could be
found within the existing political and economic
structures.[5] During the conference, an open
letter from Prince L'vov to workers urging their
cooperation had appeared in the press, but its
naive and patronizing tone only emphasized the gap
that the future head of the Provisional Government
hoped to diminish.[6]

Guchkov and Konovalov first approached the only
legal city-wide workers' organization, the
Workers' Insurance Council. But the council, which
since its election in January 1914 was dominated
by Bolsheviks, claimed that to represent workers
on the TsVPK was outside its competence.[7] It was
then that the central committee leaders decided to
embark on a full-scale election of workers'

delegates from all factories in Petrograd containing over 500 workers. The election was to be in two stages. During the first, electors would be chosen on the basis of 1 for every 1000 workers and 1 per factory at those with between 500 and 1000 workers. Then, electors would assemble to select 10 representatives to the central committee and 6 to the Petrograd district committee. In all, 219,000 workers from 101 enterprises, or approximately two-thirds of Petrograd's factory work force, were eligible to vote. This formula, drawn up by V.I. Kovalevskii of the Imperial Russian Technical Society, closely followed that adopted by the ill-fated Shidlovsky Commission of February 1905 which also provided the basis for representation in the St Petersburg Soviet of Workers' Deputies later that year.

The fact that the government permitted the election of workers to the War-Industries Committees to take place at all is at first glance somewhat puzzling. With the exception of Polivanov and possibly one or two other ministers, there was a profound scepticism within the Council of Ministers concerning Guchkov's motives and fear that the election campaign would stimulate agitation and unrest. Furthermore, the ministers had by no means renounced their intention to proceed with the militarization of labour. The Soviet historian, S.V. Tiutiukin, has hypothesized that the ministers gave their assent to the elections hoping to weaken the committees by playing on irreconcilable differences between workers and industrialists.[8] But this view presupposes a strategy that was otherwise totally lacking in the ministers' deliberations, ignores their very real concern that an alliance between liberals and workers could be effected, and, in the absence of any substantiating evidence, must be rejected. Much more probable is that the ministers felt themselves on the defensive, given the aura of patriotism which surrounded the committees and the crisis at the front, and were persuaded by Polivanov's argument that 'it would be most careless to stir up the workers by denying them participation in a guiding organization'.[9]

The election campaign, both before and after the first stage, was an unprecedented phenomenon in Russia. Never before had so many factory workers had such an opportunity to express their attitudes on the question of collaboration with the industrial bourgeoisie. Certainly not since the beginning of the war could workers so openly meet to discuss political questions.[10] The workers' parties, caught almost unawares by the opportunity, feverishly composed instructions, sent their activists among the factories, and succeeded in turning the campaign into a political event of significant proportions. An analysis of this campaign, first in Petrograd and then in other cities, is therefore in order.

The basic positions adopted by the political factions with respect to the elections in Petrograd may be briefly summarized. First, those who considered 'Prussian militarism' a greater menace than Tsarism and advocated national self-defence as well as the self-protection of the working class stood unequivocally for electing representatives to the VPKs. They included G.V. Plekhanov and other *émigré* social democrats associated with the Paris journal *Prizyv* (The Call), A.N. Potresov and other Menshevik-defensists within Russia represented by the newspaper *Rabochee Utro* (Workers' Morning) and the journal *Nashe Delo* (Our Task), plus right-wing Socialist Revolutionaries such as Kerensky.

Among those in the Menshevik-Internationalist camp, two positions emerged. The foreign secretariat of the Organizational Committee (OK) which counted among its members Iu. Martov, P. Akselrod and A. Martynov, argued for

> the widest possible arena for struggle for which all mass organizations must be used - electors to the War-Industries Committees, workers' committees to assist refugees, trade unions, cooperatives, a Soviet of Workers' Deputies, workers' congress, etc.[11]

This line of participating in both legal and underground organizations, pursued by the Menshevik Duma fraction and the Central Initiative Group in Petrograd, none the less excluded the War-Industries Committees. In February 1916 the OK explained this exception on the grounds that participation 'risks being interpreted by the masses themselves as a step towards the unity of

classes for defence of the country', a step which
it considered 'can become fatal for Social
Democracy'.[12] On the other hand, Mensheviks who
had been banished to Siberia - the so-called
Siberian Zimmerwaldists - viewed the workers'
groups in a different light. According to the
declaration drawn up for electors by F.I. Dan and
approved by I. Tsereteli, E. Broido, K.M.
Ermolaev, and the Bolshevik A.I. Golubkov,
participation would

> given an arena for organization . . . to oppose
> the preaching of 'war to the end' with the
> slogan of peace . . . and to protect the elemen-
> tary interests of workers threatened by
> exploiters who hide underneath patriotic
> phrases.[13]

The Siberian Zimmerwaldists recommended the election of workers' delegates until such time as an
all-Russian workers' congress could be convened.
Finally, the Socialist Revolutionary (SR) Inter-
nationalists and Bolsheviks adopted a 'boycottist'
position, one implacably opposed to participation.
But whereas the SRs urged workers to abstain from
the elections altogether, the Bolsheviks'
Petersburg Committee recommended in late August
that the first stage be used to form a Soviet of
Workers' Deputies in preparation for, or in
conjunction with, a general strike and an armed
uprising - in short, to repeat in abbreviated form
the pattern of 1905.[14] This scenario, however,
was considered too optimistic by the recently
revived Russian Bureau of the party's central
committee. The Bureau merely instructed electors
to make maximum use of their meetings for anti-war
revolutionary propaganda.[15]
What this summary of the factions' positions
suggests is that the pre-war dichotomy between the
revolutionary underground and those seeking an
alternative to it - the so-called 'liquidation-
ists' - persisted into the war period even while
the crucial issue of the war and how to respond to
it altered the configuration adhering to each
tendency. The most notable development in this
respect was the split among the Mensheviks, with
Martov and other members of the OK rejecting the
opportunity extended to workers to 'accumulate

their organizational forces' in favour of their
conception of internationalism. Those Mensheviks
within Russia, consisting of the Siberian exiles,
literatori concentrated in the capital, and the
'practical men' (*praktiki*) involved in trade union
and cooperative work tended to subordinate the war
issue to the opportunity to organize.

Of course the extent to which the factions'
principled positions influenced Petrograd's
factory workers, many of whom had only recently
arrived from the countryside and were experiencing
their first taste of social democratic agitation,
should not be exaggerated. In 16 factories,
according to police, the majority of workers voted
in favour of the Bolshevik 'boycottist' resolution
but at least five of these (New Lessner, Putilov,
Ericsson, Aivaz and Nobel) sent Menshevik electors
to the meeting on 27 September.[16] The reverse
also seems to have occurred. Indeed, although the
majority of resolutions adopted at factory
meetings supported participation or at least
participation conditional upon the convocation of
a workers' congress, the final vote taken by electors rejected participation by a margin of 90 to
81.

Among the factors contributing to the boycottist
victory were the prorogation of the Duma in early
September which demonstrated the weakness of the
liberal opposition movement, the wave of strikes
in Petrograd which reinforced workers' militancy,
and the hesitation of many Menshevik electors to
commit themselves before the question of a
workers' congress had been resolved. K.A. Gvozdev,
a Menshevik worker from the Ericsson Telephone
Factory and chairman of the electors' meeting,
found an additional reason. In a widely publicized
letter, he revealed that the co-chairman of the
meeting, the Bolshevik S. Ia. Bagdat'ev, had
impersonated a non-party elector from the Putilov
Works.[17] This irregularity provided the TsVPK
with another chance to obtain worker representation, and on 29 November the electors met for
the second time. The arrest of several Bolshevik
electors in the interim, a more stable situation
in the capital, and the already successful election of workers' representatives in Moscow
favoured a reversal of the earlier decision. A
walkout by Bolshevik and left-wing SR electors

assured it. Sixteen representatives - 13 Men-
sheviks and 3 SRs - were chosen. Among the 10
allotted to the central committee were Gvozdev who
acted as chairman, 3 delegates from the Petrograd
Pipe Factory, 1 each from New Lessner, Lebedev,
Obukhov, Putilov and the state-owned Gunpowder
Workers, and V.M. Abrosimov, a worker from the
Promet factory who had been recruited by the
Okhrana (secret police) before the war.[18]

The election of workers' representatives in
Petrograd was at best only a partial victory for
the advocates of conciliation and compromise
within the Russian industrial bourgeoisie. While
it raised the possibility of cooperation between
bourgeois liberals and socialist workers, it by no
means committed the latter to such a policy. Nor
did it ensure that agreements reached between the
two would be binding on their respective constitu-
encies. The fact that the workers' representatives
had agreed to associate themselves with an organization dedicated to 'war to the end' made them,
despite Dan's declaration, a concrete example of
'social chauvinism' in opposition to which the
Bolsheviks appeared as the champions of class
intransigence and internationalism. On 1 December,
the Bolsheviks' Petersburg Committee issued a
leaflet which accused 'Mr. Gvozdev and Co.' of
being 'traitors and renegades acting behind the
backs of the working class...for the honour of
lounging in the soft easychairs of the War-
Industries Committees'.[19] The police reported
shortly afterwards that workers at the New
Lessner, Nobel and Ericsson factories threatened
to cast Gvozdev and several other delegates out of
the factory grounds on wheelbarrows unless they
withdrew from the committees. At the Baranovskii
and Promet plants there were temporary work stop-
pages in protest against the election.[20]

Taking note of the tendentiousness of their
election and the hostile reaction it had provoked,
Gvozdev and his fellow 'war-industrial socialists'
defended their participation in the committees on
two grounds. The first repeated the Siberian
Zimmerwaldist position that 'the committee is
important as an opportunity to organize'. 'Send
this committee to the devil together with
Guchkov', proclaimed a delegate from the Petrograd
Pipe Factory, 'but we need this opportunity.'[21]

The purpose was neither 'to postpone our strife and struggles until after the war' as Guchkov had urged, nor to contribute directly to the war effort. It was, in the words of another Pipe Factory delegate, 'to fight for workers' rights' and 'to prevent an increase in production at the expense of workers'.[22] Hence, when the 10 representatives met the central committee for the first time on 3 December, it was to declare themselves an autonomous workers' group bearing no responsibility for the decisions of the host organization and making its legitimacy contingent upon the decision of an all-Russian workers' congress.

As far as the war was concerned, Gvozdev announced that the workers' group stood for the destruction of neither Germany nor Russia but for 'self-defence'. This position, as one historian has already pointed out, did not preclude revolutionary action.[23] For according to the workers' group, the Tsarist regime was waging 'a merciless war on its own people' which was 'bringing the country to the brink of defeat'.[24] The only way to prevent this was to overthrow Tsarism. But it was equally clear to the members of the workers' group that the working class by itself could not accomplish this task. It was too weak, and besides, it would have to wait its turn. As Gvozdev contended at the November meeting, 'we stand on the eve of a bourgeois revolution'. Thus, in line with the classic Menshevik formula, the second function of the workers' group was to 'nudge the bourgeoisie forward in its timorous struggle for power'.[25]

In the midst of its tribulations over securing the election of workers' representatives, the TsVPK began to encourage district and local committees to arrange elections of their own. Even before 29 November, campaigns had commenced in Nikolaev, Ekaterinoslav, Khar'kov, Samara, Nizhnii Novgorod, Saratov, Grozna and Krasnoiarsk, and in Moscow, Minsk, Lugansk and Kazan' they had culminated in the election of representatives.[26] The Moscow election contrasted with Petrograd's in that very few pre-election meetings were held, the campaign was short and the turnout was relatively low. Of 89,948 eligible voters in 62 enterprises only 45,987 (51 per cent) actually voted. Textile enterprises predominated in the election, reflecting the weight of that branch of industry in Moscow. This might explain the low turnout, as over half of the work force in textiles consisted of women whose degree of literacy and public activeness was relatively low. Of the 10 representatives chosen from among 91 electors (including 2 women), 6 were from textile enterprises. The chairman of the 10, V.A. Cheregorodtsev, was the only one from Moscow's metalworks industry.[27]

While the Moscow election occurred as planned, many other campaigns ran into difficulties no less serious than those encountered in Petrograd. In Nizhnii Novgorod, workers' electors, primarily from the large Sormovo Machine Works, met on 4 October and voted 28 to 6 to boycott the War-Industries Committee. After long tumultuous campaigns, electors in Khar'kov and Tiflis also voted against participation.[28] The boycottist movement was unwittingly assisted by local police officials. Where there was little or no interference by police as in Moscow, Kiev, Samara and Omsk, the committees had no trouble obtaining large majorities in favour of participation.[29] On the other hand, electors' meetings in Tiflis and Saratov resolved to disband rather than vote for representatives in the presence of police officials. In Ekaterinoslav, the regional military authority's ban on workers' gatherings, the arrest of many electors including those advocating participation, and, after the ban was lifted, the presence of police at the electors' meeting succeeded in sabotaging the formation of a workers' group. Similar bans in Tomsk, Arkhangel'sk and Kursk had a similar effect.[30]

Finally, the committees themselves often constituted a major obstacle to the participation of workers. On various pretexts committees in Kherson, Mariupol', Riazan', Simferopol' and Volynsk refused to follow the central committee's directives and did without workers' groups. The Odessa committee postponed elections until the summer of 1916 on the grounds that such an event would provoke excesses against Jews. After four workers' representatives had been elected, the committee denied them the autonomy other committees had granted and the representatives withdrew.[31]

Nevertheless, despite external provocation and the faint-heartedness of many committees, 58 workers' groups with a total membership of nearly 200 were established by 1917.[32] Many existed in small remote towns such as Sukhumi, Semiplatinsk, Khabarovsk and Kamyshin (Saratov province). But that this was the rule, as several Soviet historians have asserted, is simply not true. Workers' groups functioned in Kazan', Kiev, Moscow, Nikolaev, Perm, Petrograd, Rostov-on-Don, Samara, Taganrog and Tsaritsyn all of which were important industrial centres or ports. If the war was responsible for bringing these workers together under the aegis of a bourgeois organization, it did not necessarily reduce their opposition to the government's attempts to impose greater labour discipline, industrialists' desires for higher productivity and the maintenance of social peace, or the Bolsheviks' more sweeping vision of a proletarian revolution, in this case arising out of miseries induced by the war. Whether they were successful in using the workers' groups to resist these contending forces, that is, whether they 'skilfully exploited' their organizational framework as F. Dan was later to argue, or, as G. Zinoviev claimed, were used by the 'Tsarist committees', can best be ascertained by examining their activities in some detail.[33]

7.2   THE POLITICS OF INDUSTRIAL RELATIONS

The activities of the workers' groups can be divided into three overlapping phases, each of which was determined by, and was a response to, shifts in the forces impinging on them. During the first, the groups sought to establish a *modus operandi* with the committees and to gain legitimacy in the eyes of their constituents. In the second phase which lasted until October 1916, the groups concentrated on realizing the resolutions of the VPKs' second congress while at the same time fighting for survival, particularly in the provinces. Finally, failing to achieve their objectives, the workers' groups in Petrograd formed factory cells which attempted to mobilize workers around the slogan for a provisional revolutionary government, an effort that was very much part of the February Revolution.

Autonomy was a precondition for the existence of the workers' groups. In conformity with Konovalov's conciliatory approach, the central committee indulged its workers' group by providing office space, a monthly stipend, compensation to members for lost workdays, and access to its printing press. The workers' group took advantage of these opportunities by appointing a secretariat consisting of the Menshevik *literatori* Ev. Maevskii, B.O. Bogdanov and L.M. Pumpianskii, and by issuing a bulletin for distribution to provincial groups.[34] On the other hand, the Moscow committee, the base of Russia's ascendant national bourgeoisie, proved less pliant to the demand of its workers' group for autonomy. While Riabushinskii was as outspoken in his criticisms of the government as Guchkov or Konovalov, he and other Moscow committee members tended to view an autonomous workers' group as complicating rather than strengthening their cause. Only after three plenary sessions stretching over as many months and direct intervention by Konovalov was a compromise reached. The ten workers' representatives under Cheregorodtsev were permitted to comprise an autonomous group with its own secretariat but had to participate *en bloc* in the committee's section for the mobilization of industrial labour. This section, graced by an impressive array of liberal and left-leaning intellectuals, limited its activities to collecting information on the labour market and passing on requests for deferments to the military authorities.[35]

The achievement of autonomy may also be seen in the context of the groups' struggle to distinguish themselves as genuinely workers' organizations. This struggle had two forms. One involved publicly defending the interests of the urban poor through the performance of 'small deeds'. In January 1916 the central workers' group developed a scheme for compensating workers at enterprises forced to close due to lack of fuel or materials.[36] It also proposed the participation of consumer cooperatives in municipal food distribution and the reorganization of Petrograd's labour exchange along the lines of parity between workers' and employers' representatives. These programmes met with some success, though diminishing supplies of food in Petrograd and the imposition in June 1916 of additional restrictions on labour mobility militated against their effectiveness.[37] In Moscow the workers' group campaigned for municipal rent control and took an active part in the election of worker's representatives to the local insurance boards.[38]

The other method of winning rank and file support was the campaign for a workers' congress which the central workers' group launched at its first joint session with the TsVPK on 3 December. Both the workers' group and Konovalov envisioned a congress which would create an All-Russian Union of Workers with cells in each town and, in Konovalov's words, 'some kind of Soviet of workers' deputies' at the top.[39] The groundwork for the congress was laid by an organizational committee consisting of representatives of the Petrograd and Moscow workers' groups, cooperatives, and several Menshevik activists including, according to a police agent, Martov's brother, V. Levitskii (pseudonym for V.O. Tsederbaum), and Iu. Larin (pseudonym for M.A. Lur'e).[40] The committee decided to limit participation to workers' organizations with over 100 members, plus factory workers from all major industrial centres on the basis of 1 per 1000 workers in Petrograd and Moscow and 1 per 500 for other cities. This scheme, much discussed in and generally favoured by the liberal and legal socialist press, foundered when official approval for meetings with constituents was sought. The groups contended that such meetings were necessary to inform workers of the congress's programme, though the popularization of the groups themselves may also have been a motive. But having been advised by Polivanov that the meetings would 'take on a political character', A.N. Khvostov, the Minister of Internal Affairs, invoked the law of 4 March 1906 according to which police had to be present and speeches had to be submitted beforehand to the authorities.[41] These conditions were clearly unacceptable to the workers' groups. With a mandate from the VPKs' second congress to 'take all possible measures to convene the workers' congress', Konovalov and Gvozdev approached Khvostov's successor, Stuermer, in April 1916. The minister promised 'to think over this question' but 6 weeks later reiterated his predecessor's decision.[42] This plus a threat from the Moscow police to close down the War-Industries Committee if work proceeded on the congress and the spectre of Bolshevik delegates using the congress to attack the groups' 'liquidationism' were sufficient to prevent its convocation.[43]

The fate of the abortive workers' congress reveals much about the dilemma faced by the workers' groups and their sponsors. Determined to organize Russia's factory workers for self-defence, the groups found it impossible to demonstrate the advantages of this strategy so long as they operated within the legal framework. Their reliance on the War-Industries Committees as honest brokers predicated on the latter's political influence and economic importance further entangled them in this contradiction. The more Konovalov tried to portray the workers' groups to industrialists and state officials as a bulwark against industrial anarchy, the less chance the groups had of convincing workers that they represented their true interests. Yet to engage in underground activity was to risk confirming the suspicions of the government, the right-wing press and those industrialists who had never been completely won over by Konovalov's vision, that the VPKs were harbouring a revolutionary element. This dilemma intensified after the committee's second congress in February 1916 and constituted the most serious problem for the workers' groups during their second phase.

The congress witnessed many sharp exchanges between workers' representatives and those from the War-Industries Committees. But this did not necessarily indicate the failure of 'the liberal bourgeoisie of the Kadet-Progressist stripe' to find a 'common language with even "defensist" workers'.[44] What it did signify was a split within the liberal opposition movement over its relationship to the working class, with the VPKs – thanks to Konovalov – and certain Progressists and left Kadets occupying the left flank. Despite the acrimonious debate, the congress adopted nearly every measure proposed by the workers' caucus, thus going a good deal beyond the Progressive Bloc's programme. Instead of the Bloc's highly ambiguous demand for a government of public confidence, the general resolution signed by Konovalov referred to a Council of Ministers responsible to the Duma. Gvozdev's resolution on the food crisis spoke of the necessity for 'an immediate reorganization of power based on responsibility to the people's representatives', and this, together with the demand for the abolition of all restrictions based on religion and nationality (rather than the Bloc's call for 'further steps' in that direction), was approved by the congress. Finally, the groups' proposals for a workers' congress, conciliation boards, factory elders, a minimum wage and other measures relating to labour were incorporated into resolutions which the congress passed and which distinguish it from any other sponsored by industrialists before 1917.[45]

The congress's resolutions were in fact such a departure from the low profile assumed by the Progressive Bloc that they provoked wild charges from the right and concern from within the Bloc lest they 'break apart the social movement which up to this moment has been unified'.[46] Of course, the resolutions were just that – statements of intent depending on the cooperation of the government, industrialists and workers for their implementation. They set the stage but could not compel the actors to perform their assigned roles. As the following months demonstrated, attempts by the workers' groups and the leaders of VPKs to realize the resolutions were out of step with the march of events. The spring and summer of 1916 marked the high point of Russia's military operations. Although the supply system remained chaotic, the lull in fighting had enabled the army to re-equip itself with the result that it became less dependent on the public organizations. General Brusilov's offensive and Rumania's entry into the war on the allied side boosted the government's morale, making it less vulnerable to criticism. In such circumstances, the congress's demands for political concessions drove the government to adopt the opposite policy with serious consequences for both the committees and the workers' groups. But as illustrated by the campaigns to institute factory elders and conciliation boards and to intervene otherwise in industrial disputes, government intransigence and repression were only partly responsible for sabotaging the self-regulation of industrial relations. It is to these campaigns that we now turn.

The idea of factory elders originated as an extension of the unionization movement sponsored by Colonel Zubatov, head of the Okhrana's Moscow branch, and had actually received legal sanction. According to the law of 10 June 1903, factory workers obtaining the permission of management could choose candidates from among whom the management would select elders. Their selection was in turn subject to the approval of the provincial governor who retained the right to remove any elder at any time. This cumbersome scheme, analogous in many respects to and probably derived from the village elder system, smacked of the paternalism which characterized the Tsarist regime's approach to the labour question prior to 1905. Before it ever got off the ground, it was superseded by more independent forms of worker organization such as trade unions and soviets.[47]

The disinterment of the factory elder system by the workers' groups and the leaders of the War-Industries Committees can be seen as an attempt to find the least common denominator for workers' organization acceptable to both labour and management as well as to the government. Its very limited application would suggest that at least in the context of the First World War such common ground was not to be found. In addition to ministerial prevarication, there was widespread opposition to the election of factory elders among the industrial bourgeoisie. Both the Petrograd and Moscow Societies of Factory and Mill Owners demanded the right to supervise election meetings, arguing that 'it is not unknown that extreme elements will use any pretext for applying principles foreign and hostile to the social order which appears in their eyes as bourgeois'.[48] The Moscow VPK wanted to restrict the candidature of elders to those who were at least 25 years old and had worked in the same enterprise for at least a year. Even then it was only prepared to encourage elections at factories engaged in fulfilling orders for the committee.[49] Industrialists in other parts of the country were even less enthusiastic about the institution. In May 1916, 25 of the largest factory owners in Ekaterinoslav declared their categorical opposition to it as did the Russian Society of Shipbuilders several months later. The Urals district VPK in Ekaterinburg reported that owners were indifferent to the factory elder system, while in Perm several elders, elected *sui generis*, were dismissed from their jobs.[50]

The fact that the Perm workers took such an initiative would suggest some grass-roots support for the revival of the system. Indeed, factory elders were elected in the capital at Ericsson, Aivaz and the Petrograd Pipe Factory, in Nizhnii Novgorod at the Sormovo Machine Works, and at a few enterprises in Kiev and Lugansk.[51] But this hardly demonstrates an overwhelming response. While many older workers retained vivid memories of trade union and soviet activity, among those recently arrived from the countryside and hence familiar with (though not necessarily enamoured of) an analogous institution, a sizeable proportion were under 18 and therefore ineligible to vote. Faced with such obstacles, the factory elder system once more succumbed to its pre-war fate.

The campaign for conciliation boards, pursued simultaneously with that for factory elders, provoked even more stringent opposition. The Minister of Trade and Industry, whose revised version of a factory elder system failed to get through the Council of Ministers, argued before the Duma that conciliation boards were dangerous because their decisions in favour of workers would encourage the intransigence of their counterparts in other factories.[52] The central workers' group, anticipating the government's rejection of compulsory arbitration, proposed the formation of an optional (*fakul'tativnyi*) board to be sponsored by the TsVPK and containing equal representation from the workers' group and the Petrograd Society of Factory and Mill Owners.[53] Adopted without revision by the second congress, the proposal none the less was unacceptable to Petrograd's industrialists. Ironically and perhaps disingenuously, their major objection was that the workers' group would not be able to make any settlement stick because of the lack of a trade union network in the capital.[54]

As a result of such non-cooperation, the central committee found itself in a quandary. To designate its own representatives as the workers' group urged would have confirmed its class rather than 'public' nature and was in any case no guarantee that management would comply; to refrain from doing so, however, would jeopardize the establishment of a mechanism for arbitration and strain relations with the workers' group. Rather than resolving the dilemma, the committee persisted in trying to convince the government to draw up legislation and at the same time renewed its efforts to attract representatives from the Petrograd Society to its optional board. Neither approach succeeded and nothing more was heard about conciliation boards in the capital until after the February Revolution.

In Moscow, it was not until May 1916 that the VPK considered the implementation of the second congress's resolution. The delay was indicative of the committee's lukewarm attitude. Taking its cue from its convalescing chairman, the committee considered that sponsorship of a conciliation board without legislative approval would 'lead to a number of complications in our internal business and detract from the committee's fulfilment of its basic tasks'. In other words, those of the vice-chairman of the Moscow Society of Factory and Mill Owners, 'there are more important questions to consider than conciliation boards'.[55] In Rostov-on-Don, the district committee, noting that wage increases and labour turnover were getting out of hand, saw the conciliation board as a means of regulating both. However, it, along with the Perm and Samara committees, understandably doubted that workers would cooperate.[56] Such complications only confirmed the suspicions of the right wing *Moskovskie Vedomosti* that 'if such institutions in France and Germany are weakly developed because of unfavourable conditions, then they are not at all appropriate to Russia'.[57]

While the workers' groups persevered in their attempts to establish factory elders and conciliation boards, workers throughout the country were increasingly turning to the strike as a means of achieving their demands. The workers' groups were not opposed to the strike weapon but argued that alternative mechanisms for resolving industrial disputes ought to be developed and that in their absence, the groups themselves could perform this function. This position was based on two assumptions: first, that strikes weakened and divided the struggle of 'Russian society' against the autocracy, and second, that wartime conditions were unfavourable for the success of strikes. Thus, in the view of a member of the central workers' group, the proletariat's struggle 'does not require calls to strike, meetings before the factory gates, or the declaration of fine-sounding resolutions, but prolonged preparation'.[58]

That this position proved unpopular and ultimately untenable may be attributed to three factors. Firstly, the assumptions on which it was based were questionable. Not all of Russian society was struggling against Tsarism, though many industrialists and other bourgeois elements were critical of its policies. Hence, there was no question of dividing an already divided society. As to whether wartime conditions militated against successful strikes, workers did have the advantage that unless employers wanted to risk falling behind in fulfilling war orders and suffer the sequestration of their enterprises they had to come to terms with workers' demands. Figures compiled by M. Balabanov show that the percentage of strikes 'lost' sharply dropped in 1915 and remained less than half of pre-war levels in 1916, while those 'won' jumped from 7.2 and 5.8 per cent in 1913 and 1914 to 18.5 in 1915 and 24.0 in 1916.[59]

Secondly, the groups' position – to approve of the right to strike but to oppose strike action – was at best confusing and at worst hypocritical. The distinctions which Gvozdev drew between social peace and the regulation of conflict, and between the coordination of progressive social forces and the class struggle may have made sense to social theorists but their subtlety escaped most industrialists and workers. During a big strike at the Putilov Works in February 1916, the central workers' group issued a statement to the effect that while 'strikes are a completely legitimate form of protest', workers should immediately return to work. In May, the month which saw the highest number of strikes since the beginning of the war, the group asserted in a circular to the

Duma that

> one cannot blame the working class for the
> element of the strike but those who put the
> working class in a position where it has no
> other instrument for the protection of its
> interests except the strike.[60]

Yet, at a meeting of district committee representatives held in the same month, the members of the
workers' groups in attendance stated that they
considered strikes to be 'extremely undesirable at
this time'.[61] Thus, the groups could be accused
of strikebreaking by more militant workers and of
instigating strikes by, among others, General
Ruzskii, Commander of the Northern Front, and
General Lozino-Lozinskii, the governor of
Perm.[62]

Finally, the groups' attempts to resolve industrial disputes and intervene in strikes, even
according to the bulletins of the central workers'
group, were rarely effective. On those occasions
when striking workers requested the groups to
intervene, they were thwarted by district
committee censure, by the refusal of the factory
administration to recognize them as a legitimate
negotiating agent, or by state intervention. The
most spectacular failure in the latter respect was
the strike at the 'Naval'' shipyards in Nikolaev
in February 1916. Not only was the intervention of
Abrosimov and a representative of the central
committee fruitless, but the yards were closed by
order of the Minister of the Navy, most of the
strike leaders were arrested, and several thousand
workers were conscripted.[63] This pattern was
repeated with slightly less disastrous consequences at the New Lessner factory in April and
the Nikopol'-Mariupol' metallurgical plant in May.

Such defeats weighed heavily on the workers'
groups and the War-Industries Committees. Neither
had benefited by association with the other. If
the groups constituted an organizational base
within the working class, they were unable to
build on it. The VPKs, on the other hand, had not
succeeded in wresting control of industrial
relations from the bureaucracy and the threat of
disruptions to production remained ubiquitous. One
serious consequence of this state of affairs was a

deterioration of relations within the organization. Konovalov, an early and forthright defender of autonomy for the workers' groups, defended in May 1916 the censorship of their bulletins by the central committee. While the Moscow group complained of unwarranted interference, the central committee cancelled a banquet to honour Albert Thomas rather than allow Gvozdev to deliver an uncensored speech in his presence.[64]

At the same time, the Menshevik factions, which had previously either tolerated or turned a blind eye to the workers' groups, openly attacked them. Already in March 1916, Martov, Larin and other Menshevik-Internationalists threatened to expel from the party all those who did not resign from the central group - a threat which was ignored and came to nought.[65] But in August, the Central Initiative Group in Petrograd, which had been in close contact with Gvozdev, printed a leaflet condemning the workers' groups for their 'national chauvinism' and claiming that in lieu of their failure to hold a workers' congress they had forfeited their right to exist.[66]

Thus assaulted from right and left, the workers' groups were also subjected to a systematic campaign of harassment by the police. By late May, 6 of 12 members of the Samara workers' group were under arrest, and in June the entire workers' group in Rostov-on-Don met with the same fate.[67] From the middle of July, police raids on the headquarters of the Moscow workers' group became a frequent occurrence, and in August, the secretary, S.M. Monoszon (S.M. Schwarz), was ordered to leave the Moscow military district at short notice.[68] These acts were part of a larger effort approved by the Council of Ministers in June to intimidate and if possible silence the public organizations in general and the VPKs in particular.

The turning point for the workers' groups came in October 1916 in connection with a series of strikes in Petrograd and other cities. Meeting on 17 October, the workers' group under Gvozdev issued an appeal 'to the working population of Petrograd' which read in part:

Rumours are flying. In Moscow one hears of revolution in Khar'kov, in Khar'kov about

revolution in Moscow. Is it not possible that at
the base of these rumours there is some guiding
invisible evil will? Workers, be on guard! Given
our disorganization, *provocateurs* can divide and
weaken us.[69]

Welcomed   by   the   Kadet   party's   newspaper,
Rech',[70] and undoubtedly appreciated by liberal
circles in general, the appeal did not dissuade
over 100,000 Petrograd workers from striking and
was the last such action taken by the central
workers' group. As *Den'* reported, the decision to
oppose the strikes was by no means unanimous,
several representatives from the Moscow, Perm,
Ekaterinburg, Kazan' and Omsk workers' groups
considered resignation, and at a meeting on 23
October a speaker defending the group's position
was received 'with equanimity' while another who
shouted 'to the devil with this workers' group'
received applause.[71]

Hence, in addition to the pressures already
cited, the workers' groups faced a revolt within
their own ranks. It was at this point that the
central workers' group changed its strategy and
went on the political offensive. On 2 November, in
the wake of the strikes in Petrograd, the group
met with Menshevik deputies of the Duma to discuss
ways of 'saving the country from the government'
that would involve 'the Duma and all public,
workers' and democratic organizations'.[72] In
line with this aim, the workers' group organized
'assistance groups' (*gruppy sodeistviia*) in
several large factories, met with representatives
of 11 provincial workers' groups, and drafted a
new slogan – 'the immediate [and] decisive abolition of the existing regime and the creation of a
provisional revolutionary government in its
place'.[73] The third phase of the workers'
groups' activity, to be discussed in connection
with the challenge of revolution, had begun.

By renouncing their strategy of operating within
legal confines, the workers' groups were, in
effect, acknowledging the fruitlessness of their
previous efforts and the fallaciousness of the
assumptions on which those efforts had been based.
It implied that the problem of legitimacy which
arose in connection with their elections had never
been resolved. Restricted in their activities by

the government, their parent organization and
their own scruples, the groups had not developed a
mass following. Consequently, they lacked the
social weight necessary for reconciling the social
democratic leadership to their existence and
dealing with the VPKs from a stronger position.
The *volte face*, which marked the beginning of the
groups' third phase, was dictated as much by their
sense of isolation and the threat of liquidation
as it was by the upsurge in strike activity and
general political unrest. But the workers' groups
were not alone in reassessing and radically alter-
ing their political orientation and the actions
deemed appropriate to it. In late 1916, the
leaders of the VPKs were also preparing for a
confrontation with the Tsarist state, but in their
own way, and, unfortunately for them, in their own
time.

8 The Challenge of Revolution

8.1 FROM PATRIOTIC DESPAIR TO REVOLUTION

The twin spectres of repression from above and revolution from below haunted the VPKs' leaders no less than their associates in the liberal opposition movement. They were as outraged by the prorogation of the Duma in September 1915 as they were alarmed by the ensuing wave of strikes and workers' demonstrations. However, unlike the Progressive Bloc with which they were associated, the VPKs were not paralysed by their situation between the Tsarist devil and the deep red sea of proletarians.

Miliukov, leader of the Kadet Party, seemed permanently transfixed by what he termed the 'black-red bloc' - 'provocation from the right' and 'the incendiarism of an armed uprising'.[1] For him it was better to wait out the war than submit to these pressures. 'We must be more moderate for the time being, conserving our strength for decisive action after the war', he told delegates to the Union of Towns' fourth congress.[2] M.V. Chelnokov, chairman of the Union and a right-wing Kadet, shared Miliukov's predilections. 'We must arm ourselves with patience and wait', was his advice to a meeting of the Progressive Bloc's leaders in October 1915. After the war there would be time for 'serious discussions with Goremykin'. Guchkov disagreed. He advocated a different strategy:

> I would raise a fighting slogan and enter into direct conflict with the authorities regardless of the consequences. I would be prepared to wait for the end of the war, if a favourable outcome

were guaranteed. But we are being led to a total
defeat abroad and toward domestic ruin. The
government is defeatist....Its authority is
flabby and rotten.[3]

But the Bloc stuck to its policy of passivity,
waiting, as the Kadet V.A. Maklakov put it, for
some *deus ex machina*, perhaps another 11 March
1801.[4] Miliukov's faith in the Duma as the only
instrument of political remonstrance was unshakeable. It neatly coincided with the desire of
Chelnokov, Prince L'vov, N.I. Astrov and N.M.
Kishkin to steer the public organizations clear of
political questions.[5]

Guchkov had no such faith in the Duma. He also
had no scruples about the VPKs' involvement in
politics. But throughout the first half of 1916,
Guchkov was compelled to remain outside politics
while he recuperated from a near-fatal heart
attack. Thus, the banner which he raised in
October 1915 had to be taken up by his assistant,
Konovalov. Having done more than anyone else to
extend the liberal opposition movement to
encompass Menshevik-defensists, Konovalov now
steered the committees in a direction which not
all liberals found acceptable. This was the
political significance of the VPKs' second
congress.

The congress in fact marked a turning point for
the committees, representing as Zhukovskii noted
on the last day, 'a big shift to the left'. For
those who counted on the committees to defend
industrialists' interests much as the Association
of Industry and Trade had done before the war,
this was dangerous because, to cite Zhukovskii
again, 'the government and public opinion in
Russia are unprepared'.[6] Like the more cautious
members of the Progressive Bloc, they wondered
whether the committees, in accommodating themselves to the workers' groups, were not being
deflected from their true purpose. On the other
hand, it is probably fair to say that for
Konovalov, the left Kadet, N.V. Nekrasov, and
moderate socialists such as S.N. Prokopovich, V.G.
Groman and A.V. Peshekhonov – all of whom took an
active part in the congress – a common programme
was the congress's primary purpose.[7] If the
outcome of the congress distressed some of the

VPKs' members, the reaction among Petrograd industrialists and the regime was far more hostile. Previously content to ignore the committees, those who formed the Association of Metalworks Industry Representatives in February-March 1916 now accused the organization of having created a 'heavy atmosphere' by its investigation of Putilov, Nevskii and other enterprises, of having 'swallowed' the Association of Industry and Trade, and of leading it onto an 'unacceptable path' with respect to the labour question.[8]

The VPKs became the object of special concern for court circles soon after the February congress. On 15 March 1916, the Empress Aleksandra Fedorovna wrote to her husband 'I wish you could shut up that rotten War-Industries Committee as they simply prepare anti-dynastic questions for their meetings.'[9] First haltingly, then with unmistakeable clarity, the Council of Ministers attempted to reduce the scale of the committees' economic activities and isolate them politically. In March, G.V. Glinka, Assistant Minister of Agriculture, informed Konovalov that in view of the congress's condemnation of the government's food policy, he would no longer include the committees in any of the ministry's projects.[10] Polivanov's dismissal on 13 March was a big blow to the VPKs. Shortly thereafter, the Council of Ministers advised the military departments to bypass the committees in the distribution of large orders.[11] This was followed in June by a directive 'on the necessity of gradually reducing the distribution of orders through the committees'.[12] Despite the strenuous objections of the committees and the liberal press, the directive was executed. The disbursement by the military departments to the TsVPK markedly fell after June 1916. Orders distributed for the first half of the year totalled 105.44 million rubles, while from July through December the value of orders was only 83.54 million (see Table 5.1). As no decline in orders distributed directly to factories or through agencies such as the Vankov Organization can be detected, the VPKs' complaint of politically motivated discrimination seems to have been justified.

The Ministry of Internal Affairs, one of whose tasks was to restrict the VPKs' political activities, pursued it with particular zeal after February. The ministry developed a plan for the revision of the committees' statutes so that state officials could comprise a majority in the central committee.[13] A governors' conference, convened by Stuermer in April, heard calls for the outright liquidation of the organization but limited itself to approval of the ministry's plan.[14] Support for this course came as well from the Commander of the Moscow Military District, General I.I. Mrozovskii, who wrote to the ministry that

> the introduction of the government's representatives would be...completely natural in view of the huge sums granted by the state to these organizations and will have a restraining influence on their activities.[15]

Others in the military favoured the more drastic measure. General N.V. Ruzskii, Commander of the Northern Front, counselled the War Minister, Shuvaev, to

> concern yourself with the annihilation of the TsVPK which consists of defeatists. This committee does not serve the interests of defence but revolution. Strikes of workers at defence factories threatening us with defeat stem from it. Such a role is also played by local VPKs which have become shelters for the Yids, Armenians, and other national minorities. ...The business of defence must be in the hands of military people, and private industry working to meet the requirements of war must also be in its hands, not those of revolutionary committees.[16]

Shuvaev and several other ministers recoiled from Ruzskii's suggestion. Even the plan submitted by the Ministry of Internal Affairs seemed to go too far. The War Minister claimed that despite the VPKs' imperfections, they performed valuable services. Naked repression would not increase their efficiency but only turn them into martyrs.[17] Instead of liquidation or the revision of the committees' statutes, the Council of Ministers devised a series of measures to win away the less fervid members of the committees and

severely limit the political appeal of the more
intractable. Apart from the resolution to reduce
orders which has already been mentioned, the
ministers decided to 'spread information among the
populace on the negative aspects of the War-
Industries Committees' work', to apply judicial
measures against individual members of committees
failing to fulfil their obligations, and to
require local administrative approval for all
congresses and meetings attended by non-
members.[18] The Tsar added over his signature
that he hoped the measures would not remain a dead
letter.

In fact, the measures were already a matter of
routine policy. A survey of the public organizations' clandestine political activities compiled
by the police was an open secret. It contained
reports of numerous unofficial meetings attended
by members of the Trudovik and Social Democratic
fractions of the Duma at which strongly worded
statements were made and possibly incriminating
resolutions passed.[19] In May 1916, the Minister
of Justice, A.A. Khvostov, announced his intention
to apply penalties to the public organizations for
a variety of illegal transactions and several
weeks later, they became law.[20] The prosecution
of the Krasnoiarsk VPK for delivering shoddy boots
to the army heralded the application of the principle of collective responsibility which was later
to be used in Samara.[21] As far as congresses
were concerned, the governors' conference had
decided that they were not essential to the
performance of the public organizations' functions. On 21 May, the head of the Moscow police
banned all congresses and meetings in the city
until further notice.

But after the ministers had devised their
strategy, the campaign against the public organizations, and particularly the VPKs, was intensified. The press, informed by an anonymous
'competent person', was full of stories of alleged
violations of the public trust.[22] Procurators
were told to pay special attention to the criminal
violations of members of the VPKs.[23] On 24
August, the Moscow VPK was informed that its
Izvestiia would thenceforth be subjected to
preliminary censorship because of its 'failure to
observe the conditions for publication'. The

committee was also required to furnish a list of subscribers.[24] On 1 September, the Tsar signed new procedures for approving congresses and meetings. They provided for the presence of police at both public and closed gatherings, and thus went beyond the law of 4 March 1906 which the formulators invoked.[25]

State repression of the VPKs appeared to be having its intended effect. The Association of Industry and Trade, having suffered something of an identity crisis since the VPKs' first congress, severed all ties with the organization in April 1916.[26] At the same time, S.N. Tret'iakov, a close associate of Riabushinskii, withdrew from the Moscow committee to devote himself to the formation of an All-Russian Union of Trade and Industry. He thereby laid the groundwork for Riabushinskii's quiet exit from the VPKs and his leadership of the union in 1917. In June, Chelnokov lashed out at the VPKs' labour programme which, according to his information, workers regarded as nonsensical.[27] When the Union of Towns prepared to hold a congress on the food crisis in November, Chelnokov rejected any participation by representatives of the workers' groups or trade unions. 'Here comes Konovalov again', he jeered, 'with his new comrades to say a lot of superficial and unnecessary things.'[28]

The VPKs might have succumbed to these pressures were it not for two developments. The first was the rise of the 'third element' in the committees, those between industrialists and workers who executed the directives of the central and Moscow bureaus and performed other essential functions. As the enthusiasm of many industrialists waned and as the workers' groups were driven underground or into gaol, this element, consisting of lawyers, academics, engineers and employees of the VPKs, provided much needed cohesion. Its emergence was foreshadowed at the second congress. During the sixth session when accusations were hurled back and forth between industrialists and workers, S.S. Raetskii, an engineer from Moscow, spoke in the name of the 'middle group' for compromise. His speech was interrupted seven times by applause.[29] Other specialists soon rose to prominence in the committees. Professor A.A. Manuilov, called upon to chair the Moscow committee's labour section, championed the plan for conciliation boards and factory elders. V.N. Pereverzev, an engineer, led those opposed to the organization's pro-syndicate bias. Meetings of employees began on an irregular basis in May 1916 and culminated in the formation of a union after the February Revolution.[30]

The second development was a shift in emphasis away from concerns directly related to the war – supplies to the army, the regulation of the war economy – to the state of the economy in post-war Russia. With the reduction of orders from the military departments and the intensification of efforts to emasculate the committees politically, their plans for the demobilization and reconstruction of industry and transport gave them a new lease on life. 'Demobilization has already become the requirement of the moment', announced the central committee's *Izvestiia* in August 1916.[31] Konovalov elaborated on this slogan in his oration to Moscow industrialists a month later:

> The government has set itself the aim of refusing to allow the VPKs to exist after the war. But the government hardly imagines what awaits us at the end of the war. These committees mobilized industry and they must also demobilize it. Without the committees the demobilization of industry will become its ruin; without a definite plan for demobilization which can be drafted by the committees and no one else, this demobilization will turn into anarchy.[32]

One of the key questions involved in demobilization was how to redirect resources to peacetime purposes without causing excessive social and economic dislocation. That railway construction would constitute one of the major areas of endeavour was implicit in the 5-year programme introduced by A.A. Bublikov to a conference of VPKs in October 1916. The programme was a gloss on one drawn up by I.N. Borisov, an official in the Ministry of Ways and Communications.[33] Two considerations guided Bublikov: the necessity to supplement existing lines rather than constructing new ones, and the desirability of expanding both domestic and foreign trade. The conference brought

together representatives from 30 VPKs, many armed
with commentaries on lines proposed by Bublikov
and proposals of their own. After 3 days of
debate, a list of 52 lines covering 28,000 versts
of track was composed for further consideration by
the Council of Ministers and the legislative
organs. Another 55 lines covering 25,000 versts
were prescribed for construction after 1922.[34]

Such a programme, which would have nearly
doubled the existing length of track, begged the
question of who would provide the necessary
capital and who would exploit the newly accessible
resources. Would it be the state or private
enterprise? Would domestic or foreign capital
predominate? Under what conditions and from which
countries could Russia expect to attract foreign
capital? These and other issues were debated by a
special financial-economic commission which met
irregularly between October 1916 and February 1917
under the auspices of the Association of Industry
and Trade but with VPK members liberally represented. The contrast between the assessment of
Russia's future projected by the Association in
the early months of the war and the deliberations
of the commission is highly instructive. Then it
was forecast that Russia would be able to dictate
its own economic programme; now, given the necessity of liquidating the state's huge debts, it was
generally agreed that 'we cannot escape economic
dependence on the allied countries in one form or
another'.[35] Equally painful was the consideration of the state's economic functions in the
post-war era. Its monopolization of the production
of certain goods was considered a 'necessary evil'
by some but categorically rejected by others.[36]
The old issue of 'state socialism' was raised at
the last session on 20 February 1917 by A.A.
Vol'skii, a long-standing member of the Association and former editor of its journal. Reflecting
on the failure of the war to fulfil industrialists' expectations, Vol'skii gave an extremely
gloomy view of what Russia could expect in the
future:

> For me it is clear that three years of this war
> has brought humanity closer to socialism much
> quicker than the peacetime work of a generation.
> I view each state railway, municipal enterprise,

waterworks, lighting, cooperative, syndicate, etc. as bringing us closer to a socialist structure which will replace the capitalist structure.[37]

The significance of these discussions as far as the VPKs were concerned was twofold. First, the fact that they were held at all is evidence that industrialists would not abide by what Konovalov called 'the narrow departmental character of state planning'.[38] Secondly, the substance of the meetings was such as to give further notice of the VPKs' intention to extend their life into the post-war era.

In the meantime, however, relations between the government and the committees deteriorated still further. It was almost certain that if the former survived the war the latter would not. The opening shot in this, the final, round of conflict was Guchkov's letter of 15 August 1916 to General Alekseev, the army's Chief of Staff. The letter contained attacks on several ministers, mentioned the 'solid reputation' of Stuermer as a traitor, and warned that 'the flood is nearing'.[39] The letter, which was widely disseminated, caused a considerable stir when copies reached various ministers and through them the imperial court. Rumours that an order had been issued for Guchkov's arrest reached Moscow.[40] The Tsar let it be known that further oppositional activity on Guchkov's part would lead to his banishment from the capital.[41]

In the light of what is known about Guchkov's plot to overthrow the monarch, it is tempting to see his letter to Alekseev as part of the conspiracy.[42] It is undeniable that Guchkov was seeking Alekseev's sympathy; it is quite possible that he publicized the letter to put pressure on the general. But that he was trying to incriminate Alekseev in a seditious enterprise or even recruit him is less certain. There is no concrete evidence that the attempted *coup d'état* was already in the making.[43]

There is even less basis to suppose that the VPKs were involved at this time in such an enterprise. That the committees were suspected by the Tsar and some of his advisers of being seditious is important to acknowledge, but they are not the most authoritative sources in this matter. The aim of this study has been to demonstrate that groups working within the framework of the Tsarist regime contributed to its downfall not because of conspiracies, but because their attempts to improve on the administration of the economy and the war effort frequently duplicated and otherwise negated the state's own efforts and were in turn cancelled out by the state. Conflicts over how and who was to mobilize industry and transport *demobilized* support among that very sector of 'society' which in other countries had rallied around, and was incorporated into, their governments. Neither the bourgeoisie nor the state succeeded in bridging the gap between 'society' and the nation. Thus, the sacrifices demanded of 'society' in the name of the nation, essentially that it should cohere by avoiding class conflict, had limited appeal.

To return to Guchkov's letter, it appears more as an act of despair at the fortunes of the public organizations and Russia's military situation – which had begun to deteriorate again – than cold calculation designed to compromise the generals.[44] The VPKs' fortunes were at their low point in mid-August. Credits had been cut off since June. Congresses had been banned. Provincial workers' groups had been arrested. Guchkov felt his hands tied because as he noted in his letter,

> our methods of struggle are double-edged and can – owing to the excitable state of the popular masses and in particular of the working class – become the first spark of a conflagration, the dimensions of which no one can foresee or localize.[45]

In turning to Alekseev he was quite possibly soliciting the general's intervention to alter the government's 'disgusting policy'.[46]

The political events of the next few months are well known.[47] The government's attempt to win over accommodationist industrialists and Duma deputies reached a new stage with the appointment of A.D. Protopopov, an Octobrist deputy in the Duma, as Minister of Internal Affairs. The Progressive Bloc was at first impressed by such a concession to 'society'. Konovalov even went so

far as to predict a government with minister-
Kadets as well as Octobrists.[48] But Protopopov's
demagogic plan to take over the administration of
food distribution and suspicions aroused by his
meeting with a German banker in Stockholm quickly
dissipated the Bloc's optimism. Alarmed by the
possibility of a separate peace but emboldened by
the forthcoming session of the Duma, the Bloc
decided to launch an attack on the bureaucracy.
Miliukov had his opportunity to demonstate the
efficacy of parliamentary struggle and used it at
the first sitting of the Duma on 1 November. His
juxtaposition of the folly of Stuermer's and
Protopopov's policies with (unproven) allegations
in the press of treason was an explosive one.[49]
Nine days later, Stuermer resigned.

A.F. Trepov, Stuermer's successor, stepped up
efforts to drive a wedge between right- and left-
leaning members of the Progressive Bloc by holding
consultations with, and promising concessions to,
the former. However, the opposition movement,
inspired by Miliukov's audacity, was sustained by
the public organizations. When the authorities had
banned a joint congress on food supplies in early
November, Chelnokov and Prince L'vov argued
against taking any illegal steps in response.[50]
But a month later, in defiance of a similar ban,
they hosted in Moscow their organizations' fifth
congresses. The congresses were dispersed by the
police but managed to pass resolutions. They urged
the Duma 'to carry to the end its struggle with
this shameful regime' and called for a government
'satisfactory to all sections of the population'.[51] Chelnokov dispatched a letter to
Rodzianko, a copy of which reached the police.
'The decisive hour has come', it said. 'Every
effort must be made to form a government which,
uniting the people, will lead the country to
victory.'[52]

It was then the VPKs' turn. Representatives from
various district committees met in Petrograd under
Konovalov, and after being told by the police to
disperse, issued a resolution to

> turn to all public organizations . . . with a
> call not to lose courage and to devote all their
> strength in the common struggle for the honour
> and freedom of the country.[53]

From where did the public organizations acquire their boldness? If we accept the notion that their leaders bowed to the wishes of more militant constituents, we still must answer why at this juncture they did so and not at an earlier date. We know that by December discussions had taken place about a *coup* and how to organize it.[54] For the leaders of the public organizations, the overthrow of the Tsar or his forcible abdication were attractive alternatives to the prospects of military defeat and revolution which was expected to follow. The 'dark forces' which the congresses' resolutions claimed were ruling and ruining the country could only be removed by force. In this sense, the congresses were a *cri de guerre* as well as a *cri de coeur*.

The 'dark forces' consisted above all of the Tsar, the Empress and their beloved Rasputin. On 16 December, the latter was removed by assassination. This act, which had no connection with the public organizations or their leaders, had little or no effect on the government's policies.[55] It was then that Prince L'vov, through an intermediary, approached the Grand Duke Nikolai Nikolaevich with the idea of forcing the Tsar to abdicate. Guchkov, apparently unaware of this plot, gathered around him M.I. Tereshchenko, N.V. Nekrasov, General A.M. Krymov and Prince D.L. Viazemskii to develop a plan for waylaying the imperial train.[56] The process by which these conspiratorial links were forged has provoked much speculation. Membership in both the public organizations and resuscitated masonic lodges was common to most conspirators. There were, however, leaders of the public organizations, such as Riabushinskii and Bublikov, and masons such as Kerensky who remained outside, if not entirely ignorant of, the conspiracies.

The conspirators announced their intentions to wider circles including their associates in the public organizations, but did not ask for any assistance. P.A. Buryshkin, active in the Union of Towns, claims that Prince L'vov's plot was met with scepticism and incredulity.[57] This is confirmed by N.I. Astrov, another confidant of Prince L'vov.[58] We know that the actual number of conspirators in both plots remained small and that neither plot got out of the discussion stage.[59]

Years later in emigration, Guchkov, the only conspirator to leave memoirs, wrote

> Never in all the time of my political activity was I more sure that I was taking a step in favour of the monarchy as that moment when I took part in the conspiracy against the Tsar, Nicholas II.[60]

He explained the necessity for taking this step in a letter written in 1931:

> I hold to my former point of view that a spontaneous revolution from below could be prevented exclusively by a rational revolution from above, an operation (surgical intervention) which, having become inevitable, should have been carried out not by the butcher of the nearest butcher shop with his Russian axe but by a surgeon. The ruling classes should have taken this act upon themselves. But the street took it upon itself and what good can you expect from the street?[61]

Despite all that had occurred in the intervening years, there is no reason to doubt that these were indeed the motives which guided Guchkov and his fellow conspirators in 1916-17.

The conspiracies to overthrow the Tsar constituted one path chosen by the heads of the public organizations. They neither involved nor defined the nature of the organizations' activities on the eve of the February Revolution. Guchkov did not recruit the VPKs for the onslaught on autocracy, but the workers' groups had already begun their onslaught and the committees did not restrain them. This campaign had much more to do with 'the street', as Guchkov contemptuously referred to the strikes and demonstrations of February 1917, than the conspiracies.

Beginning in November 1916, the central workers' group attempted to organize workers in the capital for political action. The form this action was to take, as reported by police informers, was a march to the Duma to coincide with its reopening. The groups in Petrograd and Moscow also took the unprecedented step of encouraging workers to strike on 9 January 1917 in commemoration of Bloody Sunday.[62] These plans, however, were interrupted by the police. A.D. Protopopov, who as Minister of Internal Affairs had already informed the War Minister that he considered 'the further existence [of the workers' groups] as completely intolerable', received a report on 2 January which claimed that 'at the present moment the leading role in the revolutionary movement belongs to the so-called workers' group of the Central War-Industries Committee'.[63] Following the incarceration of the Moscow and Samara workers' groups, the police in Petrograd arrested all but three members of the central and Petrograd district groups on 26/27 January.[64] That the groups had intended more than simply to celebrate the reopening of the Duma was made clear by a leaflet which the police uncovered in a search of their premises. Composed on 12 January after the government had postponed the Duma's session, the leaflet proclaimed:

The working class and democracy can no longer wait. Each day that passes brings new danger. The decisive overthrow of the autocratic regime and the complete democratization of the country are now tasks requiring immediate execution.

Workers must rally their forces without delay, elect factory committees, come to an arrangement with workers of other factories, and towards the opening of the Duma prepare for a general organized demonstration. Let all workers of Petrograd simultaneously march to the Taurida, factory after factory, district after district, to proclaim the fundamental demands of the working class.

The entire country and the army must hear the voice of the working class. Only a provisional government depending on those organized in struggle can lead the country out of this impasse and fatal collapse, strengthen its political freedom, and lead it to peace on conditions acceptable to the Russian proletariat as well as the proletariat of other countries.[65]

But was the march to the Duma, now scheduled for

14 February, supposed to signal the revolution? It was certainly no mere Gaponian gesture; the workers' groups were agitating for the overthrow of the autocracy not concessions from it. On the other hand, it was to the Duma not the barricades that the masses were being directed. The proclamation of 'the fundamental demands of the working class' could not in and of itself bring down the autocracy. Rather, it was designed to bring working class pressure to bear on 'society' as embodied by the Progressive Bloc of the Duma to nudge it forward. As Ev. Maevskii, one of the formulators of this strategy, later wrote,

> We thought that on the one hand a public demonstration would appeal to the working masses, and on the other that as a petitioning movement, peaceful but with revolutionary slogans in the name of saving the country, it would meet with the sympathy of broad layers on the non-working class population.[66]

These calculations were too optimistic, for the majority of those adhering to the Bloc reacted negatively or not at all to the projected march. Miliukov announced to a meeting called by the TsVPK to discuss the arrest of the workers' groups that the Duma alone would dictate the conditions of struggle and that he would have nothing to do with any demonstrations. This he reiterated to the Kadet Party's central committee on 4-5 February and shortly afterwards, in a letter to *Rech'*, appealed to the workers of Petrograd to ignore the march.[67] Rodzianko, as chairman of the Duma, added his voice to those urging workers not to take to the streets.[68] Among those at the central committee's meeting who called for a militant response to the arrest of the workers' group, Guchkov and Konovalov were conspicuous by their silence. They subsequently defended the group as a force for moderation within the working class and agitated for the release of those arrested from prison.[69]

The differences that emerged from the meeting were symptomatic of the varied responses to the challenge of revolution. The Duma liberals wanted power but would accept it only as it devolved from the Tsar. Guchkov and other 'liberals with spurs' planned to take power forcibly from the Tsar if only to prevent 'the street' from seizing it first. Finally, those with less fear of, or contempt for, 'the street' were willing to carry on the worker's group's campaign.

On 14 February when the Duma reopened, 84,000 workers from 52 enterprises in Petrograd were on strike. To what extent the agitation of the workers' groups was responsible for this situation is far from clear. Soviet historians have claimed that workers were responding to the call of the Bolsheviks' Petersburg Committee and cite the fact that the majority of demonstrators gravitated towards Nevskii Prospekt as instructed by the Bolsheviks, while only 400 proceeded to the Tauride Palace where they were easily dispersed by police.[70] But the Petersburg Committee's call was virtually an eleventh hour one, issued to make the most of a situation which it had little part in creating. The fact that relatively few of the demonstrators marched to the Duma may be explained not only by a Bolshevik leaflet which condemned such a manoeuvre as subordinating the labour movement to the bourgeoisie, but by the prior arrest of its principal advocates as well as extraordinary precautions taken by the police which were reported in the press.

Obviously, more information is required before a definitive judgement about the events of 14 February can be made. The composition and activities of the assistance groups, for example, remains obscure. Did they consist solely of Menshevik 'defensists' or were non-party and even Bolshevik workers involved? Did they change the nature of their agitation once the members of the workers' groups had been arrested? What became of them after the demonstrations? In any case, one should not lose sight of the fact that the majority of workers participating in the strikes of 14 February were neither Mensheviks nor Bolsheviks and that in striking they were neither supporting nor opposing one or the other party faction. Just as in October 1916, doctrinal differences between 'defensists' and 'defeatists' paled before the overwhelming call for the overthrow of the Tsarist government and an end to the war. In October the workers' groups went against the tide; now they were part of it. After two

weeks of strikes, bread riots, political demonstrations and mutinies, the workers and soldiers
of Petrograd achieved their first objective.

8.2 THE WAR-INDUSTRIES COMMITTEES FROM FEBRUARY
     TO OCTOBER

Confronted with the revolutionary situation in
Petrograd, the War-Industries Committees and the
liberal opposition movement of which they were
part remained on the sidelines, presenting, in the
words of a recent account, 'a sorry spectacle' of
'self-interested restraint'.[71] On 28 February
1917 the Moscow committee's bureau held an extraordinary session at which Riabushinskii urged
caution. S.A. Smirnov, the vice-chairman of the
committee, suggested a vaguely worded resolution
calling for ministers responsible to the people
and a parliamentary system. In the end, two telegrams were dispatched, one of support to the
Duma's Executive Committee and one to provincial
VPKs asking them to continue their work without
interruption. On the following day, the central
committee did the same.[72]
    When the TsVPK's *Izvestiia* reappeared on 13
March after a hiatus of more than two weeks, its
front page contained the enormous banner headline:
'Pledge of Freedom – Unity and Victory!'. The
contributions which the committees could make
towards unity and victory were outlined in the
same issue. Having mobilized thousands of
factories and workshops for war production and
possessing a formidable technical staff, the VPKs
were directed to intensify their work to increase
production of goods for the army. At the same
time, as a public organization, the committees
were to continue their efforts to 'educate the
masses in the necessity for unity' because 'there
has never been a moment when unity was needed more
than at the present'.[73]
    In the first halcyon days after the overthrow of
the monarchy, these tasks did not appear beyond
the VPKs' means. Scores of provincial committees
sent telegrams to Petrograd announcing the
formation of provincial executive organs and the
continuation of their work and that of the factories.[74] A take-over of the Vankov Organization as

well as the Special Council of Defence's factory councils, greater credit allowances, and the restitution of the cotton section's powers to distribute orders were among the projects advanced by the Moscow committee at this time.[75] The central committee organized a special session on 8 March to honour the new ministers drawn from its ranks. More than one thousand 'solid, composed businessmen', according to the *Izvestiia*, were in attendance and the hall rang with their pledges for unity and support for the Provisional Government.[76]

But unity and the smooth operation of industry did not depend on the Provisional Government or businessmen's pledges. Already on 9 March Guchkov, now Minister of War and the Navy, wrote to the Commander in Chief, General Alekseev:

> The Provisional Government has no real authority at its disposal. Its decrees are carried out only to the extent that is permitted by the Soviet of Workers' and Soldiers' Deputies which has in its hands the most important elements of real power, such as the army, the railways, the post and telegraph....In particular, it is now possible to give only those orders which do not radically conflict with the orders of the above-named Soviet.

And Alekseev's reply was hardly consoling. 'We quite possibly will return to the hopeless situation of 1915', he wired back.[77]

Dual power was established throughout the length and breadth of the country and nothing better illustrated it than what happened to the VPKs and their workers' groups. The latter made a significant contribution to the formation of the 'controlling organs of revolutionary democracy', the soviets. It was their assistance groups, meeting with representatives of sick benefit funds and cooperatives between 23 and 25 February, which have generally been credited with reviving the call for creating the Petrograd Soviet.[78] While Guchkov, Konovalov, Tereshchenko and Manuilov took up ministerial posts in the Provisional Government, Gvozdev, B.O. Bogdanov and G.E. Breido, all of the central workers' group, were original and active members of the Petrograd Soviet's Temporary

Executive Committee.[79] While provincial VPKs sent representatives to the committees of public safety, workers' groups in Kazan', Tsaritsyn and Viatka took the initiative to set up soviets, and in Moscow and Kiev, members of the workers' groups served on the soviets' executive committees.[80]

The VPKs advocated national unity in the name of defence, but their own members were more concerned about the defence of their own particularistic interests. Technical personnel formed a Union of Engineers with regional branches. Employees of the VPKs followed suit. Industrialists, too, felt compelled to exercise the new found freedom to organize. Confronted by the movement for workers' control, the Petrograd Society of Factory and Mill Owners hammered out a programme for the protection of their property and later formed the nucleus of an All-Russian Union of Factory and Mill Owner Societies.[81] In Moscow, Riabushinskii chaired the first congress of the All-Russian Union of Trade and Industry on 19 March. His opening remarks to the congress reflected both the aspirations and the fears of Moscow manufacturers. While urging support for the Provisional Government, he noted with dismay its tendencies towards state monopolization. 'We must still pass through the stage of the development of private enterprise', he proclaimed, and cited Friedrich Engels (rather than fate) in support of his contention. On the question of where Russia was to acquire the capital necessary for its all-round industrial development, Riabushinskii insisted that 'our borders must not be wide open to foreign capital'. 'This does not mean', he added,

> that we must turn away foreign capital, but that it should not be victorious, that it should be counterposed by our own Russian capital for which it is necessary to create conditions that will enable it to originate and be developed.

Turning to the more immediate demand of workers, he asked rhetorically, 'can we or can we not afford the luxury of an eight-hour day?'[82]

The TsVPK attempted to accommodate all these groups, if not their conflicting aims, through the democratization of its membership. In preparation for the organization's third congress, the

committee assigned N.N. Pokrovskii, a former State
Comptroller, to revise the membership rules. The
plan which he drew up called for a tri-partite
division applicable to all VPKs. The central
committee was to contain 25 representatives from
industrial organizations, 25 from public and
scientific institutions, and 25 workers.[83]

The plan gave the appearance of equality but
left industrialists in a better position than
either engineers and technical personnel, who had
to share power with public officials, or workers,
who were to be lumped together with soldiers'
deputies and employees of the VPKs.[84] This was
assuming, of course, that workers' soviets would
send representatives to the committees.

The third congress of VPKs met in Moscow in mid-
May. Over 1000 delegates attended including a
respectable number of workers' representatives.[85] The presidium, however, was barred to
all but the most senior members all, or nearly
all, of whom were industrialists.[86] On the
opening day of the congress, Riabushinskii's newspaper, *Utro Rossii*, urged the delegates to leave
the 'social question' to the Minister of Labour
and to concentrate instead on the production of
supplies. It held out hope for the conversion of
the committees into a Ministry of Supplies, once
again, as it did 2 years previously, looking to
Britain for inspiration.[87]

But the social question could not be avoided and
the idea of conversion found little sympathy among
the delegates. Gloom pervaded the congress as one
speaker after another testified like witnesses in
a courtroom to the anarchic situation in industry
and transport. Guchkov, reinstated as chairman
after his resignation from the War Ministry,
opened the congress. His speech was described in
the press as that of 'an old man with a tired
spirit'.[88] Konovalov, still Minister of Trade
and Industry though only for another 2 days,
pleaded with the soviets to direct the labour
movement into 'regular channels of class
struggle'. Otherwise, he warned, they could expect
the complete paralysis of economic life. Then,

when millions of people will be without work,
without bread and without blood, when the agony
of production will seize one branch of the

economy after another, everywhere carrying with
it death and destruction, then the masses will
understand how they lured themselves into an
abyss. But it will be too late.[89]

Kozakevich cited the multitude of economic demands
by workers as a primary cause of the crisis in
industrial production, and Bublikov, speaking
about the great number of railway wagons out of
service, claimed that a fall in labour intensity
at the shops was responsible.[90]

Such sallies against workers did not go
unanswered. A.I. Kabtsan, a member of the Samara
workers' group since its inception, launched a
diatribe against the previous speeches, concluding
that they marked the beginning of the counter-
revolutionary movement against the Soviet of
Workers' Deputies.[91] Representatives from 11
workers' groups announced in a joint declaration
that they could no longer remain in the VPKs.[92]
Gloom gave way to indifference. Some of the
congress's working sessions lacked quorums and
could not pass resolutions.[93] Pokrovskii's plan
for democratization was approved, but by October
only two district committees, Samara and Rostov-
on-Don, had reported its execution.[94]

Once the chimera of unity had been revealed for
what it was, the committees lost much of their
élan. The proletariat, which was being educated by
the soviets and factory committees, had little use
for the VPKs. Industrialists, who for the most
part had given up hope of educating the masses in
unity, transferred their aspirations to all-
Russian unions, Allied generosity, or General
Kornilov.

While the political significance of the VPKs
declined after the third congress, certain economic functions retained their importance for yet
awhile longer. Shortly after the congress, the
Special Council of Defence, now under P.I.
Pal'chinskii, granted the TsVPK 15 million rubles
in working capital.[95] This was the first time
that the committee had received money from the
council which was not tied to the fulfilment of
specific orders. Other state organs, particularly
the GIU, were generous with advances. In April,
the Quartermaster Unit allocated 26.3 million
rubles for orders up to July 1917. Later, it

granted in principle an advance of 90 per cent of the cost of boots which the Moscow VPK was preparing in workshops throughout the central industrial region.[96]

Due to the backlog of unfulfilled orders for grenades and shells as well as the continuing decline in metal production, the VPKs obtained few orders from the GAU in 1917.[97] However, the wish of the Moscow committee to expand the powers of its cotton section was fulfilled. By a decree of 2 June 1917, the Ministry of Trade and Industry commissioned the cotton section to distribute all cotton yarn to mills fulfilling orders for the army. In October this right was extended to orders distributed by the Ministry of Provisions.[98]

Perhaps the most ambitious project devised by the VPKs in 1917 concerned the export of potash. Having acquired monopoly rights to market this potassium salt before the revolution, the TsVPK arranged to have large quantities sent from Tsaritsyn up the Volga to Iaroslavl', and from there to Arkhangel'sk by rail from where it was to be shipped to the United States. The foreign exchange thus obtained was to be used to purchase as many as 5 million pairs of shoes, the distribution of which the Ministry of Provisions assigned to the committee.[99]

This arrangement was never realized. The strike of railway workers in September, the October Revolution, and the ensuing difficulties in trade and sea transport nullified it. The execution of the VPKs' other responsibilities was hampered by the decline in demand for military supplies and labour disputes. Thus, the percentage of textile production designated by the army for its use was reduced from 75 before July to 30 thereafter.[100] In June, the TsVPK instructed provincial committees to cancel plans for new enterprises. Those still operated by the committees were beset with workers' demands for wage increases and control over production and distribution. The Kazan' VPK reported in July that the future of its shell factory was in doubt because, already running at a deficit of several thousand rubles per month, it could not absorb another rise in wages.[101] In September, Baron Maidel told the Special Council of Defence that unless 'repressive measures' were taken against workers at the

TsVPK's Protivogaz and 'Respirator' plants, the
factories would be closed.[102]
  With hopes for unity dashed and opportunities
for continuing war-related production doubtful,
the VPKs began to lose their most illustrious
industrialists and political activists. Konovalov
returned to the Ministry of Trade and Industry in
September. M.V. Braikevich, the chairman of the
Odessa VPK, S.A. Smirnov and S.N. Tret'iakov also
accepted positions in what was to be the last
cabinet of the Provisional Government. Guchkov
became engrossed in Kornilov's escapade even while
chairman of the TsVPK. After Kornilov's rout, he
was arrested. Soon released from prison, he made a
final appearance at the Moscow committee's session
on 9 September, urging it to prepare for the
evacuation of Petrograd.[103] Others simply
drifted away without explanation.
  On the local level, too, the exodus was apparent. Provincial committees all but stopped sending
information to Petrograd and Moscow, itself an
indication of reduced activities. The Moscow
committee made a special appeal in September to
the Provisional Government to prevent

> the annihilation and destruction of an organiz-
> ation consisting of 264 cells in the provinces
> and endowed with technical personnel whose
> experience and knowledge...can help the
> government to organize economic life.[104]

But in the same month, the central committee felt
compelled to draw up and approve 'rules on
procedures for the liquidation of district and
local VPKs'.[105]
  The 'third element', especially engineers,
remained the only group for which the committees
retained their importance. It was neither the
immediate involvement of the committees in
production and distribution nor their public
character that interested this element, but the
opportunity which they afforded for planning on an
all-Russian scale. For months, the Moscow
committee's journal, *The Productive Forces of
Russia*, carried articles on the possibility of
post-war industrial development in Russia's
hinterlands. The natural resources and productive
capacities of Western Siberia, the Volga region

and Turkestan were highlighted. When in July it became clear that the Provisional Government's Main Economic Committee lacked regional complements, the VPKs appeared likely candidates to fill the void. 'The VPKs', wrote the Moscow engineer Lapirov-Skoblo,

> now number 300 [*sic*], from Vladivostok to Petrograd, from the Caucasus to the White Sea. The application of the principles of decentralization and detailization of production, as well as the reduction in cost of articles produced can only be realized by an organization such as the VPKs.[106]

Nothing came of this suggestion, in part because the Main Economic Committee itself came to nothing. But the principle of decentralization was further developed by the committees in connection with the possible evacuation of Petrograd. An article in the August issue of *The Productive Forces of Russia* contrasted the capital, whose industry operated on foreign fuel or that transported over great distances, with other areas of the country which contained abundant fuel and materials but little industry. Why is there no metalworks development in the Urals? Why no petrochemical industry in Baku, it asked.[107]

The other pet project of the engineers was the demobilization of industry which the third congress confirmed as the principal objective of the committees after the war. 'The VPKs, having mobilized themselves for war, can with their accumulated experience and knowledge mobilize themselves for the peacetime market', asserted the military engineer, V.A. Petrov.[108] A conference on demobilization held in Moscow in early October attracted only a smattering of industrialists but large numbers of technical personnel. Among them were Lapirov-Skoblo, Petrov, Pereverzev and S.M. Zil'berberg, engineers who not long afterward gave their services to the Scientific-Technical Section of the Supreme Council of the National Economy (Vesenkha).[109]

First, however, the crossing had to be made. In July, following the violent demonstrations in the capital, the TsVPK had urged the Provisional Government to put an end to dual power. 'To be strong, the Government must not share power with anyone', the committee's resolution affirmed. It advocated a 'decisive struggle with anarchy', 'iron discipline in the army at the front and in the rear'.[110] But by October, the committee was not nearly so resolute. A belated and somewhat pathetic appeal to 'citizen-workers' told the story.

> When the old regime fell, there was hope that the industrial life of the country could revive, that conditions were adequate . . . to increase productivity to the necessary limits.
> Eight months have passed. And what has happened? Everywhere reigns anarchy, pillage, brigandage, violence and hooliganism. The productivity of labour has fallen to alarming proportions. The railways will soon be halted entirely....In the cities, factories and mills are on the brink of closing. There are no raw materials, no fuel, no credits. . . .Who would give money to Russian factories when their fate is so black?[111]

Yet, despite such a depressing present and foreboding future, the committees did not welcome the October Revolution. 'The public was mobilized and mobilized industry under the Tsarist government not to support its authority but in spite of it', stated the Moscow committee's *Izvestiia*. In such a spirit, the organization resolved to continue its work. 'The harder the conditions in the country, the more it is threatened with destruction, the more energetically we must strain every nerve to fulfil our duty until the end.'[112] An editorial in *The Productive Forces of Russia*, dated 30 November 1917, predicted that the seizure of power in Petrograd by those bent on socialist transformation of the country would lead it instead to economic anarchy. 'Russian capitalism is not yet obsolete', the article contended. Perhaps not, but the immediate prospects were not encouraging. The same issue reported that Japanese consortia were buying up mining enterprises in Siberia and the Urals, and an assessment of the oil industry concluded that:

> Regardless of political conditions, it would be

necessary to give up our most precious natural resources to foreigners....The prospect for Russia is economic slavery. Its immense natural wealth will fall into the hands of foreign capital.[113]

The VPKs had almost come full circle. Created as a challenge to the Tsarist government, they continued after the February Revolution as an adjunct of the Provisional Government and resolved to carry on despite the Council of People's Commissars. They sought to adapt Russian industry to the war with the hope of making it flourish, if not during the war then in the post-war era. After the February Revolution which freed Russia from the Tsarist yoke, they assumed the further development of capitalism. By October, however, it seemed that Russian industry was destined to be carved up among foreign predators. Essentially a nationalist movement which claimed to represent society, the VPKs could not reverse the state's growing dependence on foreign loans nor the growing antipathy of the masses towards that dependence and the war. The war had indeed betrayed Russian industrialists. In 1915 they perceived it as a stimulus for economic development 'in the European manner'; by late 1917 they were fighting a desperate and losing battle against the revolutionary disintegration of industry which the war had unleashed.

Conclusion

In considering Russia's industrial mobilization
during the First World War, it can be concluded
that politics and economics were so intertwined
that neither proved to be the decisive or more
fundamental factor. The politics of industrial
mobilization were shaped in important ways by the
general level of, and contradictions within,
Russia's economic development. But the nature and
direction of that development cannot be properly
understood without reference to the political
system. Similarly, the ways and degree to which
political conflicts, generated by industrial
mobilization, were resolved profoundly affected
which sections of the population, regions and
industries would bear the brunt of sacrifices or
reap the profits of war.

The basic argument of this study is that many
conflicts over industrial mobilization remained
unresolved, such that by 1917 the authority of the
Tsarist government had been seriously undermined,
while the War-Industries Committees and other
organizations claiming to represent 'society' had
failed to establish their legitimacy within it.
What seems to have accounted for this stalemate
was the contradictory relationship between the two
forces. The state depended on the VPKs to initiate
a variety of coordinating and administrative functions and to arrange for the production of war
materials by enterprises unfamiliar with military
department specifications. On the other hand, many
state and military officials suspected the aims of
the committees' leaders and resented the intrusion
of the organization into matters traditionally
within the competence of the bureaucracy.

For their part, the committees were created in

the spirit of patriotism to assist the war effort and relied on the state for their lifeblood, orders and advances. However, the alternative which they posed to the perpetuation and extension of state bureaucratic administration was the expansion of the political and economic power of 'society'. This took various forms, ranging from support for the Progressive Bloc's campaign for a government of public confidence and the more radical variant of a government responsible to the Duma, to the self-regulation of industry, the sponsorship of workers' groups, and projects for post-war economic development. In each case, the issue was not so much the substance of the VPKs' proposed solutions, but whether the state would agree, or be compelled, to concede to them the necessary power.

The less cooperative the state proved to be, the less effective were the committees in mobilizing industry for war-related production, and the more they appeared to be a self-interested, politically motivated movement. But, by denying a substantial section of the industrial bourgeoisie access to the power it sought, the state was denying itself a good deal of support for its prosecution of the war. Its treatment of the Duma was no different. By late 1916, the only contributions which both the Duma and the public organizations could make were negative, consisting of defiance of state regulations and allegations of bad faith, stupidity and worse.

The ambiguity of the VPKs' relationship with the Tsarist state was compounded by the lack of unanimity within each camp. Initially, the committees could count on the support of the War Minister, Polivanov, and the forbearance of several other high ranking state and military officials. For a time, they were the beneficiaries of interministerial rivalries and were able to take advantage of interstices within the administration of the war economy. Even after their second congress, which confirmed the worst suspicions of their opponents and provoked a concerted campaign of discrimination and repression against them and their workers' groups, their representatives continued to sit on numerous state regulatory agencies and the flow of orders and advances, though diminished and interrupted, did not cease.

Similarly within the committees there was no lack of disunity. This had many causes. From the start, tension existed between the central committee, dominated by the 'organization men' of the Association of Industry and Trade, and the more politically active Moscow manufacturers. Partly because of the reluctance of financial-heavy industrial circles – the core of the Association's constituency – to throw their weight behind the new organization and stemming in part from the nationalist appeal which the Muscovites cultivated among provincial industrialists and the technical intelligentsia, the tension was resolved in favour of the latter faction. Nevertheless, the central committee continued to depend on the services of the Association's men, its practice of bypassing district committees in the distribution of orders remained a sore point within the organization, and the influence of the syndicates and trusts in connection with the committees' position on the fuel and raw materials questions never disappeared.

But undoubtedly the most divisive issue within the VPKs and the one that further alienated them from the state and the Petrograd-based financier-industrialists was their sponsorship of workers' groups. In seeking to extend the liberal oppositional movement to incorporate working class aspirations and to ensure social peace, the committees' leaders may, in fact, have contributed to the opposite tendencies. Industrialists who had joined the committees to obtain orders and even those who applauded resolutions critical of the government were, in general, far less enthusiastic about the programme for improving industrial relations devised by Konovalov and the workers' groups and, still less, about the latter's social democratic rhetoric. The resolutions calling for conciliation boards and other forms of arbitration which the second congress approved, though not without difficulty, marked the limit of their tolerance. The subsequent attempts to put these resolutions into practice proved unsuccessful largely because of the government's intractability, but also because of the indifference and opposition of industrialists both within and outside the organization. With little to show for having seized the opportunity to organize and with their liquidation imminent, the workers' groups radically revised their strategy. Beginning in November 1916, they embarked on a course which was to involve them in political strikes, demonstrations, and, in February-March 1917, the formation of soviets. On the other hand, the committees supplied several ministers for, and served as an instrument of, the Provisional Government.

This study began on a comparative note, and it seems appropriate to conclude on one as well. The war threw up a myriad of problems for each of the Great Powers, among which how to mobilize industry was one of the most complex. Russia's war-industrial capacity was relatively small at the outset, while the needs of its armed forces were enormous. That it accomplished much in the way of supplying its multi-million man army is undeniable. But neither the state nor those who took issue with its policies were strengthened by those achievements. There never was anything in Russia approaching the British Ministry of Munitions or the various war boards created in Germany. The War-Industries Committees were neither the Russian equivalent of their German namesake nor the Federation of British Industries which was also founded during the war. They did not unite Russian industry, but represented a new national element within it, challenging but not superseding the state bureaucratic structure and its financial-industrial allies.

This was not a matter of irresponsibility or treason, as each side accused the other both before and after the revolution. The state was not 'blind', as Riabushinskii had alleged shortly before the war; it was structurally and ideologically incapable of legitimating national bourgeois interests. Neither were the people 'orphaned'. They became increasingly restless with the rule of their 'Little Father', and, having brought about his downfall, refused to accept the bourgeoisie as their guardians.

Epilogue

The VPKs did not survive for very long after the October Revolution. At a session of the TsVPK on 4 December, N.N. Iznar, acting chairman since Guchkov's departure, was replaced by the lawyer, M.S. Margulies. The committee attempted to realize its tripartite plan for representation at this session, but a delegation from the Petrograd Soviet insisted that in accordance with the 'recent events' all public and governmental bodies had to be reorganized with workers in the majority. The session on 23 January 1918 was the first at which this principle was achieved.[1] By that time, Vesenkha had placed the VPKs under its committee on demobilization and renamed them National-Industries Committees.[2]

The war, as far as Russia was concerned, was over. To adjust to the new situation, the committees held a fourth congress in March 1918. The congress drew up new statutes which Vesenkha approved on 13 April.[3] The statutes gave workers' representatives half the seats on all committees. Representatives from industrialists' organizations, if indeed there were any left, were excluded, though individual industrialists could offer their services as 'specialists'.[4] The statutes also charged the committees with the establishment of a council of experts, whose functions Vesenkha had already outlined in its decree of 28 February. The council was conceived as a miniature planning agency. It was to 'develop a programme for the productive activity of various branches of industry, transport, agriculture and commerce as well as measures for the programme's realization'.[5] Within a short while, over 500 'experts' from more than 30 scientific and

213

technical organizations had joined the council. By
May it had begun work in earnest with the drawing
up of a plan for the irrigation of Turkestan.[6]
During the nationalization drive in June-July
1918, the council was transferred to Vesenkha's
direct care. On 24 July, with the civil war and
remobilization already at full pitch, the
committees themselves were decreed out of existence.[7] Margulies and the Odessan Braikevich
attempted to revive the VPKs in Kiev in early
1919, but this was already part of another mobilization effort, that of the Whites, which ended in
defeat.[8]

Notes

NOTES TO THE PREFACE AND ACKNOWLEDGEMENTS

1   See for example, V.P. Semennikov, *Monarkhiia pered krusheniem, 1914-1917* (Moscow and Leningrad, 1927). The bulletins of the workers' groups were published in 'K istorii Gvozdevshchiny', ed. I.A. Menitskii, *Krasnyi arkhiv* (hereafter *KA*), no. 67 (1934) pp. 28-92. See also 'K istorii "rabochei gruppy" pri tsentral'nom voenno-promyshlennom komitete', ed. I.A. Menitskii, *KA*, no. 57 (1933) pp. 43-82.

2   See N.I. Astrov and Paul P. Gronsky, *The War and the Russian Government* (New Haven, Conn., 1929), and S. Zagorsky, *State Control of Industry in Russia during the War* (New Haven, Conn., 1928).

3   A.P. Pogrebinskii, 'Voenno-promyshlennye komitety', *Istoricheskie zapiski* (hereafter *IZ*), vol. XI (1941) pp. 160-200.

4   Ibid., pp. 167 and 177.

5   N.I. Razumovskaia, 'Tsentral'nyi voenno-promyshlennyi komitet' (candidate's dissertation, 1947, deposited in GBIL).

6   Some of Sidorov's articles were republished posthumously in *Ekonomicheskoe polozhenie Rossii v gody pervoi mirovoi voiny* (Moscow, 1973). For others see Bibliography. See also K.N. Tarnovskii, *Formirovanie gosudarstvenno-monopolisticheskogo kapitalizma v Rossii v gody pervoi mirovoi voiny* (Moscow, 1958).

7   V.Ia. Laverychev, *Po tu storonu barrikad* (Moscow, 1967), and V.S. Diakin, *Russkaia burzhuaziia i tsarizm v gody pervoi mirovoi voiny, 1914-1917* (Leningrad, 1967). Diakin has

recently turned to a slightly earlier period, that of Stolypin's rule. See *Samoderzhavie, burzhuaziia i dvorianstvo v 1907-1911 gg.* (Leningrad, 1978).

8   See R.A. Roosa, 'Russian Industrialists and "State Socialism", 1906-17', *Soviet Studies*, vol. XXIII (1972) pp. 395-417; James White, 'Moscow, Petersburg and the Russian Industrialists', *Soviet Studies*, vol. XXIV (1973) pp. 414-20; and R.A. Roosa, '"United" Russian Industry', *Soviet Studies*, vol. XXIV (1973) pp. 421-5. It is White (p. 419) who views the Central War-Industries Committee as 'the chief battleground'.

9   Raymond Pearson, *The Russian Moderates and the Crisis of Tsarism, 1914-1917* (New York, 1977), esp. chs 3 and 4.

10 Norman Stone, *The Eastern Front, 1914-1917* (London, 1975) p. 14.

11 Ibid., pp. 194-211.

12 A.I. Guchkov, the only leading figure in the committees to have left memoirs, scarcely mentions the organization which he served for two years as chairman in the version published in *Poslednie novosti* (Paris) Aug-Sept 1936.

NOTES TO CHAPTER 1:   THE RUSSIAN INDUSTRIAL BOURGEOISIE

1   There has been little work on this important question which, in view of the often cited conflict between the Ministry of Internal Affairs (MVD) and the Ministry of Finance, is surprising. For two penetrating studies of the MVD see Daniel T. Orlovsky, 'High Officials in the Ministry of Internal Affairs, 1855-1881', and Don Karl Rowney, 'Organizational Change and Social Adaptation: the Pre-Revolutionary Ministry of Internal Affairs', in *Russian Officialdom: The Bureaucratization of Russian Society from the Seventeenth to the Twentieth Century*, eds Walter M. Pintner and Don Karl Rowney (Chapel Hill, N.C., 1980) pp. 250-82, 283-315. Rowney claims that 'the relatively privileged legal-social category of "noble" managed to retain a position of dominance in the central higher civil service as a whole

...' but admits that 'the simple denomination of "noble" concealed a great amount of variation...' (pp. 302-3). Significantly, the War Ministry encompassed both modes thereby perpetuating intra-ministerial struggles. See P.A. Zaionchkovskii, *Samoderzhavie i russkaia armiia na rubezhe xix-xx stoletii, 1881-1903* (Moscow, 1973) esp. chs 2, 3 and 6.

2   One is reminded of that contrast which Antonio Gramsci drew between the West where 'there was a proper relation between State and civil society', and Russia where 'the State was everything, civil society was primordial and gelatinous' (A. Gramsci, *Selections from the Prison Notebooks*, eds Quintin Hoare and Geoffrey N. Smith (London, 1971) p. 238). See also Perry Anderson, *Lineages of the Absolutist State* (London, 1974) pp. 195-235, 328-60 for a broad historical argument along these lines.

3   *Rezoliutsii Vysochaishe razreshennogo torgovo-promyshlennogo s"ezda obshchestva dlia sodeistviia russkoi promyshlennosti i torgovli v Moskve v iiule 1882 g.* (St Petersburg, 1882). For earlier congresses see E.S. Lur'e, *Organizatsiia organizatsii torgovo-promyshlennykh interesov v Rossii* (St Petersburg, 1913) pp. 75-9.

4   D.I. Mendeleev, *Problemy ekonomicheskogo razvitiia Rossii* (Moscow, 1960) p. 157.

5   Ibid., p. 138.

6   See Robert W. Tolf, *The Russian Rockefellers: The Saga of the Nobel Family and the Russian Oil Industry* (Stanford, Calif., 1976) and John P. McKay, *Pioneers for Profit: Foreign Entrepreneurship and Russian Industrialization, 1885-1913* (Chicago, 1970). On the role of Witte, the classic study is Theodore Von Laue, *Sergei Witte and the Industrialization of Russia* (New York, 1963). For a forceful but not entirely convincing critique of Von Laue's views, see I.F. Gindin, 'Russia's Industrialization under Capitalism as seen by Theodor Von Laue', *Soviet Studies in History*, vol. VII (1972) pp. 1-54.

7   McKay, *Pioneers for Profit*, p. 33.

8   The literature on these two syndicates is extensive. On their origins and the importance

of the French see Olga Crisp, *Studies in the Russian Economy before 1914* (London, 1976) pp. 159-66, 174-82; M.Ia. Gefter, 'Tsarizm i monopolisticheskii kapital v metallurgii Iuga Rossii', *IZ*, vol. XLIII (1953) pp. 75-82; René Girault, *Emprunts russes et Investissements français en Russie, 1887-1914* (Paris, 1973) pp. 364-71; A.L. Tsukernik, *Sindikat 'Prodamet'* (Moscow, 1959) pp. 37-42; P.V. Volobuev, 'Iz istorii sindikata "Proudgol'"', *IZ*, vol. LVIII (1956) pp. 107-44.

9 P.V. Volobuev, 'Politika proizvodstva ugol'nykh i neftianykh monopolii v Rossii nakanune pervoi mirovoi voiny', *Vestnik Moskovskogo universiteta, seriia ix*, no. 1 (1956) p. 72.

10 For a summary of the interesting debate among Soviet historians on whether finance capital and the Tsarist state were moving towards a coalescence (*srashchivanie*) of interests or the subjection (*podchinenie*) of the latter to the former, see K.N. Tarnovskii, *Sovetskaia istoriografiia Rossiiskogo imperializma* (Moscow, 1964) pp. 30-44, 126-9, 170-87. See also contributions by I.F. Gindin, A.P. Pogrebinskii, V.I. Bovykin and K.N. Tarnovskii in *Ob osobennostiiakh imperializma v Rossii* (Moscow, 1963) pp. 86-123, 124-48, 250-313, 419-38. On the shift towards passive investment cf. McKay, *Pioneers for Profit*, pp. 368-78, and Girault, *Emprunts russes*, pp. 511-40.

11 McKay, *Pioneers for Profit*, pp. 192-200; Olga Crisp, 'Labour and Industrialisation in Russia', in *The Cambridge Economic History of Europe*, vol. VII, part 2 (London, 1978) pp. 385-6.

12 See, for example, Ia.A. Livshin, 'Predstavitel'nye organizatsii krupnoi burzhuazii v Rossii v kontse XIX-nachale XX vv.', *Istoriia SSSR*, no. 2 (1959) pp. 110-17; R.A. Roosa, 'Russian Industrialists Look to the Future: Thoughts on Economic Development, 1906-1917', in *Essays in Russian and Soviet History*, ed. J.S. Curtiss (Leiden, 1963) pp. 193-218; Roosa, 'Russian Industrialists and "State Socialism"', pp. 395-417; Carl Goldberg, 'The Association of Industry and Trade, 1906-17: the Successes and Failure of Russia's

Organized Businessmen' (unpublished Ph.D. dissertation, University of Michigan, 1974).

13 V.T. Bill, *The Forgotten Class: The Russian Bourgeoisie to 1900* (New York, 1959) pp. 15-35; I.F. Gindin, 'Russkaia burzhuaziia v period kapitalizma, ee razvitie i osobennosti', *Istoriia SSSR*, no. 2 (1963) pp. 61-3; V.K. Yatsunsky, 'The Industrial Revolution in Russia', in *Russian Economic Development from Peter the Great to Stalin*, ed. Wm L. Blackwell (New York, 1974) pp. 114-17; Wm L. Blackwell, 'The Old Believers and the Rise of Private Industrial Enterprise in Early Nineteenth Century Moscow', in ibid., pp. 139-58; and M.C. Kaser, 'Russian Entrepreneurship', in *The Cambridge Economic History of Europe*, vol. VII, part 2, pp. 444-7.

14 A.G. Rashin, *Formirovanie rabochego klassa Rossii* (Moscow, 1958) p. 200.

15 Ibid., pp. 198-201.

16 Jo Ann Ruckman, 'The Business Elite of Moscow: a Social Inquiry' (unpublished Ph.D. dissertation, Northern Illinois University, 1975) pp. 47-8, 112-13. Ruckman argues that the same nineteen families 'recur again and again' in the annals of Moscow's economic life. See also Roger Portal, 'Muscovite Industrialists: the Cotton Sector (1861-1914)', in *Russian Economic Development*, pp. 161-96.

17 The life and times of 'calico Moscow' is vividly portrayed in P.A. Buryshkin, *Moskva kupecheskaia* (New York, 1954). See also Ruckman, 'The Business Elite of Moscow', pp. 41-5; and Thomas C. Owen, *Capitalism and Politics in Russia: A Social History of the Moscow Merchants, 1855-1905* (Cambridge, 1981) pp. 145ff.

18 'Ispoved' nabolevshego sredstva', *Promyshlennyi mir*, no. 46 (1 Oct. 1900) p. 1071.

19 Iu.B. Solov'ev, 'Protivorechiia v praviashchem lagere Rossii po voprosu ob inostrannykh kapitalakh v gody pervogo promyshlennogo pod''ema', in *Iz istorii imperializma v Rossii* (Moscow, 1959) pp. 382-3.

20 McKay, *Pioneers for Profit*, p. 293. McKay cites as the source of this 'piquant phrase', C. Ernest Dawn, 'Arab Islam in the Modern Age', *Middle East Journal*, vol. XIX (1965)

pp. 442-3.
21 N.K. Krestovnikov, *Semeinaia khronika Krest-
    ovnikovykh*, 3 vols (Moscow, 1903-4) vol. III,
    p. 115 (emphasis mine). Such yearnings were
    typical of what Owen (*Capitalism and Politics
    in Russia*, p. 208) calls 'violently xenophobic
    polemics of the merchant-slavophile alliance'.
22 Laverychev, *Po tu storonu barrikad*, pp. 30-1.
23 Quoted in E.D. Chermenskii, *Burzhuaziia i
    tsarizm v revoliutsii 1905-1907 gg.* (Moscow
    and Leningrad, 1939) p. 65. This statement,
    published in the Moscow newspaper, *Russkie
    Vedomosti*, served as the basis for numerous
    others submitted by provincial organizations.
    See V.Ia. Laverychev, *Tsarizm i rabochii
    vopros v Rossii, 1861-1917 gg.* (Moscow, 1972)
    p. 192.
24 Laverychev, *Po tu storonu barrikad*, p. 47.
25 On the economic discussions see ibid., pp. 66-
    9, 73-4, and James L. West, 'The Moscow
    Progressists: Russian   Industrialists   in
    Liberal   Politics,   1905-1914'   (unpublished
    Ph.D.   dissertation,   Princeton   University,
    1975) pp. 233-9.
26 *Utro Rossii*, no. 149, 18 May 1910, p. 1;
    ibid., no. 231, 9 Oct. 1911, p. 3.
27 Ibid., no. 153, 3 July 1911, p. 5; ibid., no.
    149, 18 May 1910, p. 1.
28 Ibid., no. 150, 19 May 1910, p. 1.
29 Ibid., no. 78, 4 Apr. 1912, p. 1.
30 Laverychev, *Po tu storonu barrikad*, pp. 80,
    92; and *Shestoi ocherednoi s"ezd predstavi-
    telei promyshlennosti i torgovli* (St Peters-
    burg, 1912) p. 28.
31 On the Riabushinskii family's economic endeav-
    ours see M.L. Lavigne, 'Le plan de Mihajl
    Rjabusinskij: un   projet   de   concentration
    industrielle en 1916', *Cahiers du monde russe
    et soviétique*, no. 1 (1964) pp. 90-104. For
    the original Russian and commentary see 'K
    istorii   kontserna   br.   Riabushinskikh',   ed.
    I.F. Gindin, in *Materialy po istorii SSSR*, 7
    vols (Moscow, 1952-9) vol. VI, pp. 603-40.
    Also, Buryshkin, *Moskva kupecheskaia*, pp. 189-
    93.
32 M.M. Tsvibak, *Iz istorii kapitalizma v Rossii:
    khlopchatobumazhnaia promyshlennost'* (Lenin-
    grad, 1925) pp. 55, 67. On the Russian Export

Company see V.Ia. Laverychev, 'Protsess monopolizatsii khlopchatobumazhnoi promyshlennosti Rossii (1900-1914 gody)', *Voprosy istorii*, no. 2 (1960) p. 143.

33 See below p. 240, n. 27. On the Neo-Slav movement, see Hugh Seton Watson, *The Russian Empire 1801-1917* (Oxford, 1967) pp. 665, 688-9.

34 'Predislovie', *Velikaia Rossiia, sbornik statei po voennym i obshchestvennym voprosam*, 2 vols (Moscow, 1910-11) vol. I, pp. 2, 5.

35 P.B. Struve, 'Ekonomicheskie problemy Velikoi Rossii', in ibid., vol. II, pp. 153-4.

36 L.N. Iasnopol'skii, 'Finansy Rossii i ikh podgotovka k voine', in ibid., vol. II, p. 112. Iasnopol'skii did not mention foreign markets but others did in such contributions as 'Russia as a Great Power', 'Russia in the Far East', and 'Colonizers of the Far East'. See vol. I, pp. 21-86, 195-216, 217-36.

37 *Utro Rossii*, no. 245, 25 Oct. 1911, p. 1.

38 For the former, so-called 'optimist' view, see Arthur Mendel, 'Peasant and Worker on the Eve of the First World War', *Slavic Review*, vol. XXIV (1965) pp. 23-33, and the restatement of his argument, 'On Interpreting the Fate of Imperial Russia', in *Russia under the Last Tsar*, ed. T.G. Stavrou (Minneapolis, 1969) pp. 13-41; for the latter see Leopold Haimson, 'The Problem of Social Stability in Urban Russia, 1905-1917', *Slavic Review*, vol. XXIII (1964) pp. 619-40, and vol. XXIV (1965) pp. 1-22.

39 This distinction is acknowledged by P.I. Liashchenko, 'Iz istorii monopolii v Rossii', *IZ*, vol. XX (1946) pp. 160-1.

40 Sidorov, *Ekonomicheskoe polozhenie*, p. 503; *Monopolisticheskii kapital v neftianoi promyshlennosti Rossii, 1883-1914, Dokumenty i materialy* (Moscow and Leningrad, 1961) pp. 501-11, 598-608, 737-8.

41 *Narodnoe khoziaistvo v 1913 godu* (Petrograd, 1914) p. 305.

42 Tsukernik, *Sindikat 'Prodamet'*, pp. 183-8. On the fuel famine see Volobuev, 'Politika proizvodstva ugol'nykh i neftianykh monopolii', pp. 71-115; M.Ia. Gefter, 'Toplivno-neftianoi golod v Rossii i ekonomicheskaia politika tret'ei iiunskoi monarkhii', *IZ*, vol. LXXXIII

(1969) pp. 76-122.
43 See speech by S.I. Timashev, the Minster of Trade and Industry, in Gosudarstvennaia Duma, IV Sozyv, *Stenograficheskie otchety* (St Petersburg, 1913) sessiia 1, ch. 2, zasedanie 31 (22 Mar. 1913) cols 52-6; also, M.Ia. Gefter, 'Tsarizm i zakonodatel'noe "reguli-rovanie" deiatel'nosti sindikatov i trestov v Rossii nakanune pervoi mirovoi voiny', *IZ*, vol. LIV (1954) pp. 170-93. On the prosecution of Produgol', the seriousness of which Soviet historians tend to minimize, see 'Produgol'', *KA*, no. 18 (1926) pp. 131-45, and T.D. Krupina, 'K voprosu o vzaimootnosheniiakh tsarskogo pravitel'stva s monopoliiami', *IZ*, vol. LVII (1956) pp. 166-74.
44 Ugroza chastnomu khoziaistvu, *PT*, no. 7 (151), 1 Apr. 1914, pp. 357-60.
45 *Zhurnal zasedanii vos'mogo ocherednogo s"ezda predstavitelei promyshlennosti i torgovli, sostoiavshegosia 2, 3, i 4 maia 1914 g.* (Petrograd, 1915) pp. 2, 29-30, 57, 87.
46 Ibid., pp. 110, 114-15.
47 Ibid., p. 113.
48 Ibid., pp. 100-1.
49 West, 'The Moscow Progressists', pp. 447-58.
50 S.G. Strumilin, *Ocherki ekonomicheskoi istorii Rossii* (Moscow, 1960) p. 537.
51 Gos. Duma, IV Sozyv, *Sten. otchety*, ses. 1, zas. 2 (12 Dec. 1912) cols 658-60.
52 Cf. Haimson, 'The Problem of Social Stability', pp. 4-8 and I.S. Rozental', 'Russkii liberalizm nakanune pervoi mirovoi voiny i taktika bol'shevikov', *Istoriia SSSR*, no. 6 (1971) pp. 52-70. For published documents relating to the Information Committee and its negotiations with the Bolsheviks, see *Istoricheskii Arkhiv*, no. 6 (1958) pp. 8-10, 12-13, and no. 2 (1959) pp. 13-16.

NOTES TO CHAPTER 2: RUSSIAN INDUSTRIALISTS AND THE INITIAL MOBILIZATION

1 Sir Bernard Pares Papers, School of Slavonic and East European Studies (London) box 7, pp. 1-12.
2 *Utro Rossii*, 20 July 1914, p. 1, quoted in

Pearson, *The Russian Moderates*, p. 16.

3   Quoted in *'Obzor pechati'*, *PT*, no. 15 (159), 1
    Aug. 1914, pp. 115–17.

4   *Khoziaistvennaia zhizn' i ekonomicheskoe polo-
    zhenie naseleniia Rossii za pervoi deviat'
    mesiatsev voiny* (Petrograd, 1916) p. 61.

5   'Predvaritel'nye svedeniia o gornoi i gornoza-
    vodskoi promyshlennosti Iuga Rossii za 1914 i
    1913 gg.', *Gornozavodskoe delo*, no. 3, 22 Jan.
    1915, p. 10283.

6   'Zhurnal zasedaniia mekhanicheskogo otdela'
    (24 July 1914), Tsentral'nyi Gosudarstvennyi
    Istoricheskii Arkhiv (hereafter TsGIA), fond
    150, opis' 1, delo 94, list' 3obv.

7   In a recent study of the 'soliders' revolt' in
    1917, Allan Wildman has concluded that 'one
    can say with reasonable confidence that the
    peasant-soliders entered the world war with
    the conviction that it was an alien enter-
    prise, the patriotic outpourings of cultured
    society notwithstanding' – Allan K. Wildman,
    *The End of the Russian Imperial Army: The Old
    Army and the Soldiers' Revolt (March–April,
    1917)* (Princeton, N.J., 1980) p. 374.

8   For a superb account of working conditions
    among Donets miners during the war, see Iu.I.
    Kir'ianov, *Rabochie Iuga Rossii, 1914-fevral'
    1917 g.* (Moscow, 1917) esp. pp. 35–7, 54–106.

9   According to factory inspectors' reports,
    there were only 70 strikes in Russia in the
    latter half of 1914. See I.I. Mints,
    'Revoliutsionnaia bor'ba proletariata Rossii v
    1914–1916 godakh', *Voprosy istorii*, no. 11
    (1959) p. 59. For the general strike in Baku
    see A.N. Guliev, *Bakinskii proletariat v gody
    novogo revoliutsionnogo pod"ema* (Baku, 1963)
    pp. 164ff., and the brief summary in Ronald G.
    Suny, *The Baku Commune 1917-1918: Class and
    Nationality in the Russian Revolution* (Prince-
    ton, N.J., 1972) pp. 52–8; and in Petersburg,
    E.E. Kruze, 'Rabochie Peterburga v gody novogo
    revoliutsionnogo pod"eme' in *Istoriia
    rabochikh Leningrada*, 2 vols (Leningrad, 1972)
    vol. I, pp. 449–60.

10 'Voina i eksport', *PT*, no. 18 (162) 15 Sept.
    1914, pp. 272–4.

11 'Frantsuzskii ekonomist ob ekonomicheskoe i
    finansovoe polozhenie Rossii, *PT*, no. 21

(165), 1 Nov. 1914, p. 406. For original French see R.G. Levy, 'La situation économique et financière de la Russie', *Revue des deux Mondes*, no. 24, 1 Nov. 1914, pp. 30–50.

12 See especially I.S. Bliokh, *Budushchaia voina v tekhnicheskom, ekonomicheskom, i politicheskom otnosheniiakh*, 5 vols (St Petersburg, 1898) vol. IV, p. 6. The author was a Polish-Jewish banker and a self-avowed pacifist.

13 'Ekonomicheskie zadachi mirovoi voiny i russkaia promyshlennost',' *PT*, no. 18 (162), 15 Sept. 1914, pp. 258–9.

14 'Tsargrad i prolivy', *PT*, no. 21 (165), 1 Nov. 1914, pp. 393–7.

15 'Evropeiskaia voina i promyshlennost',' *Neftianoe delo*, no. 15, 8 Aug. 1914, p. 3.

16 'Chemu dolzhna nauchit nas evropeiskaia voina?', *Neftianoe delo*, no. 17, 9 Sept. 1914, pp. 8, 11–12.

17 'Voina i russkaia promyshlennost', *Izvestiia Obshchestva zavodchikov i fabrikantov Moskovskogo promyshlennogo raiona*, no. 8 (Aug. 1914) pp. 17, 21.

18 'Kopiia dokladnoi zapiski', Tsentral'nyi Gosudarstvennyi Voenno-Istoricheskii Arkhiv (hereafter TsGVIA), f. 369, op. 16, d. 20, 1.2.

19 A.A. Manikovskii, *Boevoe snabzhenie russkoi armii v mirovuiu voinu*, 2nd edn, 2 vols (Moscow, 1930–2) vol. I, p. 124.

20 N. Golovine, *The Russian Army in the World War* (New Haven, Conn., 1931) p. 130; D.A. Kovalenko, *Oboronnaia promyshlennost' sovetskoi Rossii v 1918–1920 gg.* (Moscow, 1970) p. 25.

21 Golovine, *The Russian Army*, p. 132; A.L. Sidorov, 'K voprosu o stroitel'stve kazennykh voennykh zavodov v Rossii v gody pervoi mirovoi voiny', *IZ*, vol. LIV (1954) p. 165.

22 Golovine, *The Russian Army*, p. 132.

23 General B. Gourko, *War and Revolution in Russia, 1914–1917* (New York, 1919) p. 125.

24 Sidorov, *Ekonomicheskoe polozhenie*, p. 10.

25 Golovine, *The Russian Army*, p. 32; E.Z. Barsukov, *Podgotovka Rossii k voine v artilleriiskom otnoshenii* (Moscow and Leningrad, 1926) p. 114.

26 Golovine, *The Russian Army*, p. 32. At least twice, in 1906 and 1909, the recommendation to

switch from the 8- to 6-gun battery was put
forward. Only in November 1914 did the army
adopt it.

27 Tarnovskii, *Formirovanie*, p. 16. P. Karatygin,
Obshchie osnovye mobilizatsii promyshlennosti
(Moscow, 1925) p. 22; Golovine, *The Russian
Army*, p. 146.

28 Zaionchkovskii, *Samoderzhavie i russkaia
armiia*, pp. 252-65.

29 Kovalenko, *Oboronnaia promyshlennost'*, pp. 17-
18; E.R. Goldstein, 'Military Aspects of
Russian Industrialization' (unpublished Ph.D.
dissertation, Case Western Reserve University,
Cleveland, Ohio, 1971) pp. 54-98. Sidorov,
Ekonomicheskoe polozhenie, pp. 121-2 describes
the Petersburg factory as a 'workshop...
incapable of independently preparing artillery
systems' and the Perm factory as desperately
short of funds by 1914.

30 See M. Mitel'man, B. Glebov and A. Ul'ianskii,
Istoriia Putilovskogo zavoda, 1801-1917, 2nd
edn (Moscow and Leningrad, 1941) pp. 133-4.

31 (Ministerstvo torgovli i promyshlennosti)
Otchet otdela promyshlennosti za 1911 g. (St
Petersburg, 1911) p. 147.

32 A.N. Kuropatkin, *The Russian Army and the
Japanese War*, trans. A.B. Lindsay (New York,
1909) p. 308.

33 R. Girault, 'Finances internationales et
relations internationales (à propos des usines
Poutiloff)', *Revue d'historie moderne et contemporaine*, vol. XIII (1966) p. 220.

34 See V.I. Bovykin, 'Banki i voennaia promyshlennost' Rossii nakanune pervoi mirovoi
voiny', *IZ*, vol. LXIV (1959) pp. 82-135; Ia.
I. Livshin, *Monopolii v ekonomike Rossii*
(Moscow, 1961) pp. 65-70.

35 On the Vickers proposal which finally came to
fruition in 1916 in the form of a factory at
Tsaritsyn, see E.R. Goldstein, 'Vickers
Limited and the Tsarist Regime', *Slavonic and
East European Review*, vol. LVIII (1980) pp.
564-70, and Sidorov, *Ekonomicheskoe polozhenie*, pp. 122-3, 132-3.

36 Stone, *The Eastern Front*, pp. 28-9.

37 This is convincingly argued in K.F. Shatsillo,
'O disproportsii v razvitii vooruzhennykh sil
Rossii nakanune pervoi mirovoi voiny (1906-

1914 gg.)', *IZ*, vol. LXXXIII (1969) pp. 123-36.

38   The War Minister was ex-officio chairman of the War Council which had to approve all expenditures on orders. Stone claims that the council was 'an institution composed of... extraordinarily aged generals'. See *The Eastern Front*, p. 149.

39   Sukhomlinov's stories were published in the 1890s under the pseudonym Ostap Bondarenko. Stone, *The Eastern Front*, pp. 24-36, 194-8, attempts to reverse historians' negative verdict on Sukhomlinov but also mentions his involvement in some less than ethical practices.

40   V.A. Sukhomlinov, *Vospominaniia* (Berlin, 1924) p. 336.

41   A.A. Polivanov, *Iz dnevnikov i vospominanii po dolzhnosti voennogo ministra i ego pomoshchnika*, ed. A.M. Zaionchkovskii (Moscow, 1924) p. 148. According to Bruce Lockhart, the British consul in Moscow, Polivanov alleged that the Grand Duke managed to obtain a commission on all orders placed with the Putilov Works for his mistress, the notorious ballerina, Kshesinskaia. See Public Record Office (London), Foreign Office Papers, Series 371, vol. 2745, N92149: 'Lockhart enclosed in Ohindley (for Ambassador) to Grey', 12 Apr. 1916. Sukhomlinov makes the same accusation in *Vospominaniia*, p. 262.

42   For a defence of the GAU's policies in the pre-war era, see A. Bart, 'Na fronte artilleriiskogo snabzheniia', *Byloe*, vol. V (XXXIII) (1925) pp. 188-219.

43   The article is reprinted in *Documents of Russian History, 1914-1917*, ed. Frank A. Golder (Gloucester, Mass., 1964) pp. 190-2.

44   'Perepiska V.A. Sukhomlinova s N.N. Ianushkevichem', *KA*, no. 1 (1922) p. 253. By October the shortage of shells had become his 'main gigantic nightmare', ibid., p. 258.

45   Ibid., pp. 247-8, letters of 9 and 11 Sept. 1914.

46   Sidorov, *Ekonomicheskoe polozhenie*, pp. 27-9, based on report of Kuz'min-Karavaev to Lukomskii, 20 Oct. 1914 in TsGVIA, f. 29. The sixteen consisted of five state-owned fac-

tories (Obukhov, Izhev, Izhorsk, Zlatoustov and Baltic); six private firms located ih Petrograd (Putilov Works, Nevskii Shipbuilding and Engineering, Petrograd Metal, 'Phoenix', 'Westinghouse', and the Russian Company for the Production of Shells), two in Warsaw (Lilpop, Rau & Levenstein, and Vulcan), and one each in Nizhnii Novgorod (Sormovo Steel and Machine Works), Nikolaev (Nikolaev Shipbuilding and Engineering Works, known as 'Naval''), and Abo, Finland (Abo Ironfoundry and Engineering Works).

47 J.F. Godfrey, 'Bureaucracy, Industry, and Politics in France during the First World War: a Study of Some Interrelationships' (submitted for Oxford University D.Phil. thesis, 1971) pp. 3, 7. See also *Voprosy promyshlennosti i finansy*, no. 1 (8) (Jan. 1916) pp. 6-7.

48 These factors are given excessive weight by Stone, *The Eastern Front*, pp. 152-62.

49 Sidorov, *Ekonomicheskoe polozhenie*, p. 364.

50 Ibid., p. 27, citing TsGVIA, f. 962, op.2, d. 37, 1. 14-15.

51 Ibid., pp. 253-6; for an elaborate counter-factual statement see David Lloyd George, *War Memoirs*, 6 vols (London, 1933-8) vol. I, pp. 458-77.

52 For orders from Britain see D.S. Babichev, 'Deiatel'nost' russkogo pravitel'stvennogo komitete v Londone v gody pervoi mirovoi voiny', *IZ*, vol. LVII (1956) p. 285, and Kovalenko, *Oboronnaia promyshlennost'*, p. 53. For those from the United States and Canada see National Archives Division (Washington, D.C.), Record Group 261 (Russian Government Supply Committee in the United States), L37 'Otchet o deiatel'nost' russkogo zagotovitel'-nogo komiteta v Amerike i ego likvidatsionnoi komissii', 4 vols in 2 parts, vol. IV, pp. 3-4. This anonymously written report is a fascinating study of Russian-American financial and commercial relations during the First World War as well as being an indispensable guide to the material contained in Record Group 261.

53 Stone, *The Eastern Front*, p. 157; Sidorov, *Ekonomicheskoe polozhenie*, p. 598.

54 Testimony of General Smyslovskii quoted in

Manikovskii, *Boevoe snabzhenie*, vol. II, pp. 34-5. The Supreme Commission was appointed by the Council of Ministers in August 1915, largely in response to the crescendo of criticism over supply policies under Sukhomlinov. It was chaired by General N.P. Petrov and contained members of both legislative chambers. See Sidorov, 'Bor'ba s krizisom vooruzheniia russkoi armii v 1915-1916 gg.', *Istoricheskii zhurnal*, no. 10-11 (1944) p. 46.

55 Manikovskii, *Boevoe snabzhenie*, vol. II, p. 32.

56 'Eksport kazennykh deneg za granitsu', *PT*, no. 24, 15 Dec. 1909, p. 642, and 'Gosudarstvennaia oborona i promyshlennost', ibid., no. 8 (114), 15 Sept. 1912, pp. 196-8.

57 'O merakh po obespecheniiu svoevremennogo vypolneniia zakazov na oboronu i obespecheniiu armii boevogo snariadiia' (12 Jan. 1915), TsGIA, f. 1276, op. 11, d. 814, l. 1-3 obv.

58 T.D. Krupina, 'Politicheskii krizis 1915 g. i sozdanie Osobogo soveshchaniia po oborone', *IZ*, vol. LXXXIII (1969) p. 60.

59 'Morskoi Ministr Goremykinu' (25 Jan. 1915), TsGIA, f. 1276, op. 11, d. 814, l. 5.

60 'Ministr torgovli i promyshlennosti Goremykinu o dokladnoi zapiske' (23 Feb. 1915), TsGIA, f. 1276, op. 11, d. 814, l. 12.

61 This was in cooperation with the Commission on the Preparation of Explosives, established in February 1915. For details of the commission's work see V.N. Ipat'ev, *Zhizn' odnogo khimika, vospominaniia*, 2 vols (New York, 1945) vol. I, p. 440ff.

62 In April 1915 the British Ambassador, George Buchanan, reported to London that 'Grand Duke Serge blamed Vickers severely for the way they carried out their contract' (FO 371 2447 N42913: Buchanan to Grey, 13 Apr. 1915). The Special Commission's opposition to placing additional orders with British, American and Canadian firms led to friction with the British War Ministry and possibly was the cause of its liquidation. Cf. FO 371 2454 N110274: 'Memorandum on steps taken by the British Government to assist the Russian Government in procuring supplies', Kitchener to Grey, 11 Aug. 1915.

63 Sidorov, *Ekonomicheskoe polozhenie*, p. 42, based on figures in the archive of the Petrov Investigation Commission, TsGVIA, f. 962. The increase in Headquarters' demand was reflected in the monthly requirements for explosives which were 60,000 puds in Jan. 1915, but 165,000 in June and 250,000 shortly thereafter. (See Ipat'ev, *Zhizn' odnogo khimika*, vol. I, pp. 428-30). Production rose from 6300 puds in February 1915 to 100,900 in March 1916 according to Ipat'ev, *Rabota khimicheskoi promyshlennosti na oboronu vo vremia voiny* (Petrograd, 1920) p. 5.

64 E.Z. Barsukov, 'Grazhdanskaia promyshlennost' v boevom snabzhenii armii', *Voina i revoliutsiia*, no. 10 (1928) pp. 16-17.

65 A.P. Pogrebinskii, 'Sindikat "Prodamet" v gody pervoi mirovoi voiny 1914-1917', *Voprosy istorii*, no. 10 (1958) p. 28. Golovine, in *The Russian Army*, p. 152, makes a similar charge with respect to metalworks enterprises. This probably refers to an investigation of the Kolomna Machine Works carried out by a group appointed by the Special Council of Defence's Supervisory Commission. The investigators characterized Vankov's order for 40,000 shells as 'harmful...to the production of other no less essential items' (see 'Doklad komissii dlia obsledovaniia deiatel'nosti Kolomenskogo, Sormovogo, i Kulebakskogo zavodov', TsGVIA, f. 369, op. 4, d. 20, l. 189). Manikovskii, *Boevoe snabzhenie*, vol. II, p. 49, holds the Special Council and the War Minister responsible for such 'one-sided development'.

66 *Istoriia organizatsii upolnomochennogo po zagotovleniiu snariadov po frantsuzskomu obraztsu general-maiora S.N. Vankova, 1915-1918* (Moscow, 1918) pp. 207-8.

67 M.V. Rodzianko, 'Krushenie imperii', *Arkhiv russkoi revoliutsii* (hereafter *ARR*), vol. XVII (Berlin, 1926) p. 89.

68 Text of letter in *Padenie tsarskogo rezhima*, ed. P.E. Shchegolev, 7 vols (Moscow and Leningrad, 1924-7) vol. V, p. 204.

69 Rodzianko, 'Krushenie imperii', pp. 90-1.

70 B. Pares, *The Fall of the Russian Monarchy* (New York, 1961) p. 230. Cf. also his *Day by Day with the Russian Army* (London, 1915)

pp. 201–34.
71 Golovine, *The Russian Army*, p. 220.
72 Quoted in You. Danilov, *La Russie dans la guerre mondiale, 1914–1917* (Paris, 1927) p. 416.
73 Rodzianko, 'Krushenie imperii', p. 92.
74 Ibid.
75 Pares, *Fall of the Russian Monarchy*, p. 243. See Pares Papers, box 10, notes of 10 Apr. 1935 interview with Guchkov. Guchkov denies that he had anything to do with the formation of the Special Council. Pares evidently did not take Guchkov at his word and relies on Rodzianko ('Krushenie imperii', p. 92), who includes Guchkov among those present at Headquarters.
76 *Zhurnaly Osobogo soveshchaniia dlia obsuzhdeniia i ob"edineniia meropriiatii po oborone gosudarstva (Osoboe soveshchanie po oborone gosudarstva), 1915–1918 gg.* (hereafter *ZhOSO*) (Moscow, 1975) part 1, pp. 1–2. The changing membership of the Special Council is conveniently summarized in a table in Krupina, 'Politicheskii krizis', p. 65.
77 Sidorov, 'Bor'ba s krizisom vooruzheniia', p. 42.
78 The eleven were A.A. Davidov, K.P. Fedorov, N.D. Lesenko, A.P. Meller, A.P. Meshcherskii, M.S. Plotnikov, N.E. Ponafidin, A.I. Putilov, Ia.I. Utin, A.I. Vyshnegradskii and S.F. Zlokazov. None of them took part in the War-Industries Committees. For their ties with industry and finance cf. I.F. Gindin and L.E. Shepelev, 'Bankovskie monopolii v Rossii', *IZ*, vol. LXVI (1960) pp. 58–9, 62–3, 68–75; and V.I. Bovykin and K.F. Shatsillo, 'Lichnye unii v tiazheloi promyshlennosti nakanune pervoi mirovoi voiny', *Vestnik Moskovskogo universiteta, seriia ix*, no. 1 (1962) Tables 1–9. The banks and the relevant officials are: Russian-Asian Bank – Putilov (chair.); Petrograd International Commercial Bank – Vyshnegradskii (dir.); Petrograd Private Commercial Bank – Davidov (chair.); Petrograd Savings and Loan Bank – Utin (chair.). In terms of their fixed assets as of Jan. 1914, the banks ranked 1, 2, 8, 9 respectively (see I.F. Gindin, *Russkie kommercheskie banki* (Moscow, 1948)

p. 381).

79 'Osobyi zhurnal Soveta ministrov' (29 May 1915), TsGIA, f. 1276, op. 11, d. 888, 1. 272 obv. Sukhomlinov's affirmation is to be contrasted with the extremely pessimistic remarks which Putilov made to the French ambassador several days earlier. See Maurice Paleologue, *An Ambassador's Memoirs*, 3 vols (London, 1923-5) vol. I, p. 349.

80 *ZhOSO*, part 1, 'Primechaniia', p. 540; no. 4, 27 May 1915, and no. 6, 6 June 1915, ibid., pp. 27, 47, and Sidorov, 'Bor'ba s krizisom vooruzheniia', p. 45.

NOTES TO CHAPTER 3: MOSCOW TO THE RESCUE

1 See Stone, *The Eastern Front*, pp. 165ff. Stone characterizes the munitions shortage as 'a mere technical translation of the great social convulsion within Russia'.

2 Lloyd George, *War Memoirs*, vol. I, pp. 112, 159.

3 'Krizis ekonomicheskoi politiki', *PT*, no. 7 (174), 1 Apr. 1915, p. 340.

4 *Russkie Vedomosti*, no. 119, 26 May 1915, p. 1.

5 *Den'*, no. 142, 26 May 1915, p. 2.

6 *Kommercheskii telegraf*, no. 637, 4 May 1915, p. 1; ibid., no. 630, 24 Apr. 1915, p. 2.

7 *Utro Rossii*, no. 143, 26 May 1915, p. 1.

8 Ibid., pp. 1-2.

9 See, for example, 'Pochemu my tak zabotimsia ob imushchestvennykh interesakh nemtsev?' *Novoe Vremia*, no. 13864, 16 Oct. 1914, p. 4; 'Germanskie zavody v Rossii', ibid., no. 13944, 6 Jan. 1915, p. 5. Moscow made a big issue of German property although (or perhaps because) more German capital was invested in Petrograd.

10 The Kharitonov Commission's recommendations, first issued in 1910, were strongly supported by the Association, but the Council of Ministers did not act on them. P.A. Kharitonov was the State Comptroller. See *Promyshlennost' i torgovlia v zakonodatel'nykh uchrezhdeniiakh, 1909-1912 gg.* (St Petersburg, 1913) pp. 69-87. See Zhukovskii's report in *Russkie Vedomosti*, no. 111, 16 May 1915, p. 2. Litvinov-

Falinkskii's speech is in *Utro Rossii*, no. 129, 12 May 1915, p. 2.

11 *Birzhevye Vedomosti*, no. 14869, 28 May 1915, p. 3.

12 The full text of the speech is in TsGIA, f. 32, op. 1, d. 34, ll. 116-19. A censored version may be found in *Gornozavodskoe delo*, no. 25, 22 June 1915, pp. 11272-3.

13 *Den'*, no. 93, 6 Apr. 1915, p. 3.

14 See p. 220, n. 31. On the Riabushinskiis' automobile enterprise (AMO) see S.V. Voronkova, 'Stroitel'stvo avtomobil'nykh zavodov v Rossii v gody pervoi mirovoi voiny (1914-1917 gg.)', *IZ*, vol. LXXV (1965) pp. 159-65. The AMO plant was later renamed Zavod imeni Stalina (ZIS) and later still, Zavod imeni Lenina (ZIL).

15 *Russkie Vedomosti*, no. 121, 28 May 1915, p. 3.

16 *Rech'*, no. 144, 28 May 1915, p. 4.

17 *PT*, no. 11 (179), 1 June 1915, pp. 543-4.

18 O. Brandt, *Die deutsche Industrie im Kriege* (Berlin, 1915) pp. 240-1.

19 Ia. Bukshpan, 'Voennyi komitet nemetskoi promyshlennosti', *PT*, no. 11 (179), 1 June 1915, pp. 582-5. For an excellent study of German industry during the war see G.D. Feldman, *Army, Industry, and Labor in Germany, 1914-1918* (Princeton, N.J., 1966).

20 Lloyd George, *War Memoirs*, vol. I, p. 165. Also S.J. Hurwitz, *State Intervention in Great Britain: A Study of Economic Control and Social Response, 1914-1919* (New York, 1949) pp. 149-50.

21 Lloyd George, *War Memoirs*, vol. I, p. 147.

22 'K voprosu ob uchrezhdenii Khar'kovskogo VPKa', *Gornozavodskoe delo*, no. 22, 8 June 1915, pp. 11151-2.

23 *Birzhevye Vedomosti*, no. 14967, 16 July 1915, p. 5; *Trudy s"ezda predstavitelei voenno-promyshlennykh komitetov, 25-27 iiulia 1915 g.* (Petrograd, 1915) pp. 279-88.

24 *PT*, no. 29 (196), 7 Nov. 1915, p. 571; *Predstavitel'stvo obshchestvennykh grupp v voenno-promyshlennykh komitetakh* (Petrograd, 1916) pp. 5-6.

25 *Izvestiia tsentral'nogo voenno-promyshlennogo komiteta*, no. 15, 13 Oct. 1915, p. 4 (hereafter *Izv. TsVPKa*); *Izvestiia Moskovskogo*

voenno-promyshlennogo komiteta, no. 11 (Dec. 1915) p. 29 (hereafter *Izv. MVPKa*); ibid., no. 15-16 (Feb. 1916) p. 135.

26 M.P. Petrov, 'Goroda i snabzhenie armii', *Gorodskoe delo*, no. 13-14 (July 1915) p. 738.
27 *Utro Rossii*, no. 166, 18 June 1915, p. 3; ibid., no. 172, 24 June 1915, p. 3.
28 *Den'*, no. 237, 29 Aug. 1915, p. 6.
29 'A. Mastaev v *Nash golos* (Samara)' (10 Nov. 1915), Tsentral'nyi Gosudarstvennyi Arkhiv Oktiabr'skoi Revoliutsii (hereafter TsGAOR), f. 102 (1915) d. 347, l. 104.
30 *Den'*, no. 238, 30 Aug. 1915, p. 5.
31 'Doklad vtoromu s"ezdu (chlen TsVPK ot Orenburg-Turgai VPKa, P. Kobozev)', TsGVIA, f. 13251, op. 1, d. 8, l. 32.
32 *Izv. MVPKa*, no. 11, Dec. 1915, p. 54 (emphasis mine).
33 *Russkoe Slovo*, no. 281, 6 Dec 1916, p. 5.
34 TsGAOR, f. 102 (1915) d. 347 prilozhenie, ll. 222-3.
35 *Izv. MVPKa*, no. 34, Apr. 1917, p. 19.
36 Ibid., no. 31-2, Oct.-Nov. 1916, p. 53.
37 *Izv. TsVPKa*, no. 223, 4 May 1917, p. 3.
38 *An Outline of Activities of the Central War Industrial Committee of Russia* (New York, 1918) p. 3.
39 But just as the VPKs employed the services of engineers, so did the Zemstvo and Town Unions, largely devoted to food and medical relief work, recruit large numbers of agronomists, doctors, nurses and medical students. See table cited in Wm E. Gleason, 'The All-Russian Union of Towns and the All-Russian Union of Zemstvos in World War I: 1914-1917' (Indiana University Ph.D. dissertation, 1972) p. 28, showing that of 650 Union of Towns members, doctors comprised the largest group (24 per cent). Although neither Union recruited workers as members, at least the Union of Towns had a wider constituency than the VPKs among middle class groups. Especially in 1917 the VPKs and the Zemgor organization debated which was the more broadly-based. See *Izv. MVPKa*, no. 35, June 1917, p. 3; 'O klassovoi prirode Zemgora', *Izvestiia Zemgora*, no. 6, 20 May 1917, p. 2.
40 For this expression and an excellent summary

of the 'tsarist technostructure', see Kendall
E. Bailes, *Technology and Society under Lenin
and Stalin: Origins of the Soviet Technical
Intelligentsia, 1917-1941* (Princeton, N.J.,
1978) pp. 19-43.

41 *Ob organizatsii osoboi komissii po promyshlennosti pri I.R.T. obshchestve i ob ee zadachakh*
(Petrograd, 1914) pp. 1-5.

42 *Trudy komissii po promyshlennosti v sviazi s
voinoi*, no. 11, 2 June 1915. See resolution on
p. 46.

43 *Lichnyi sostav voenno-promyshlennykh komitetov*
(Petrograd, 1915) pp. 36-8, 41-2.

44 *Kievskii   oblastnyi   voenno-promyshlennyi
komitet* (Kiev, 1916) pp. 1-21.

45 Based on information for 18 district and 44
local committees provided by *Lichnyi sostav
VPK-ov*, *passim* and scattered newspaper
accounts.

46 A.D. Bilimovich, 'Finansovoe otnoshenie VPK-
ov', *Izv. TsVPKa*, no. 48, 13 Jan. 1916, p. 3.

47 *Lichnyi sostav VPK-ov*, pp. 77-83; *Den'*, no.
354, 24 Dec. 1916, p. 6.

48 *Den'*, no. 252, 13 Sept. 1916, p. 3; *Lichnyi
sostav VPK-ov*, p. 90.

49 *Birzhevye Vedomosti*, no. 14839, 28 May 1915,
p. 3; *Den'*, no. 144, 28 May 1915, p. 3.

50 *Russkie Vedomosti*, no. 124, 31 May 1915, p. 3.

51 *Gornozavodskoe delo*, no. 28, 15 July 1915, p.
11417.

52 'Zhurnaly zasedanii TsVPKa', TsGVIA, f. 13251,
op. 40, d. 30, ll. 1-168; *Izv. TsVPKa*, no.
152, 1 Oct. 1916; ibid., no. 161, 25 Oct.
1916; no. 182, 17 Dec. 1916.

53 *Den'*, no. 163, 16 June 1915, p. 3; *Izv.
TsVPKa*, no. 1, 24 Aug. 1915, p. 2.

54 *Den'*, no. 251, 12 Sept. 1915, p. 4. The resolution which the session adopted was extremely
ambiguous. Krasin, of course, had been a
Bolshevik, but his relations with the party
during the war and up to 1917 were non-
existent.

55 Kutler, the last chairman of the Association
of Industry and Trade and its candidate to the
Constituent Assembly, was arrested by the
Bolsheviks but later released. In the early
1920s he was a member of the board of Gosbank.
Maidel's books on bridge and rail construction

were published in the twenties and thirties. Iznar worked in the Commissariat of Transport.

56 On 10 June Sukhomlinov wired his approval of the three representatives from the TsVPK (see 'Sukhomlinov Avdakomu', TsGVIA, f. 369, op. 1, d. 72, 1. 23). At least two were present from 24 June onwards. *ZhOSO*, part 1, pp. 80ff.

57 *Birzhevye Vedomosti*, no. 14879, 2 June 1915, p. 2.

58 The Petrograd Society of Factory and Mill Owners established a district VPK largely, one suspects, as a pre-emptive move. It was to become the Fronde of the VPK movement. TsGIA, f. 150, op. 1, d. 60, 11. 235-78, contains minutes of the Petrograd Society's meetings at which the formation of the district VPK and issues related to the movement were discussed. The file of the Petrograd district VPK in the Leningrad *oblast'* archives was unavailable to me.

59 FO 371 2452 N77065: 'Buchanan to Grey enclosing Lockhart', 31 May/13 June 1915. See also R.H.B. Lockhart, *The Two Revolutions* (London, 1967) pp. 55-6.

60 FO 371 2454 N1106039/105582: '17/30 June 1915'. The eyewitness was William Peters, a student of Pares' at the time.

61 See P. Milyukov, *Political Memoirs, 1905-1917*, ed. A.P. Mendel (Ann Arbor, Mich., 1967) p. 322.

62 *Birzhevye Vedomosti*, no. 14888, 7 June 1915, p. 2. Prince L'vov's remarks were strikingly similar to Lloyd George's speeches to Midlands' factory workers as reported in ibid., no. 14877, 1 June 1915, p. 1.

63 T. Riha, 'Milyukov and the Progressive Bloc in 1915: a Study in Last Chance Politics', *Journal of Modern History*, vol. XXXII (1960) p. 18. Also, 'Kadety v dni Galitsiiskogo razgroma', ed. N. Lapin, *KA*, no. 54 (1933) pp. 117-22.

64 Laverychev, *Po tu storonu barrikad*, pp. 114-15.

65 *Finansovyi otchet s nachala deiatel'nosti Moskovskogo voenno-promyshlennogo komiteta po 1 avgusta 1916* (Moscow, 1917) pp. 49-52.

66 *Utro Rossii*, no. 194, 16 July 1915, p. 4.

67 'Voenno-promyshlennyi zavod v Moskve (doklad

organizatsionno-ispolnitel'noi komissii)',
Izv. MVPKa, no. 2, Sept. 1915, pp. 9-11. A.I.
Kuznetsov, chairman of the shell section, and
St.P. Riabushinskii were board members of the
company.

68 *Izv. Ob-va fab. i zav. Mosk. prom. raiona*, no.
7, July 1915, p. 13.

69 *Deiatel'nost' Moskovskogo oblastnogo voenno-
promyshlennogo komiteta i ego otdelov po 31
ianvaria 1916 g.* (Moscow, 1917) p. iii.

70 *Promyshlennaia Rossiia*, no. 607, 31 May 1915,
p. 8.

71 *Trudy pervogo s"ezda predstavitelei metallo-
obrabativaiushchei promyshlennosti, 29 fevra-
lia-1 marta 1916 g.* (Petrograd, 1916) p. 39.

72 *Zhurnaly TsVPKa*, no. 3, 9 June 1915. A.P.
Korelin, 'Monopolii v metalloobrabativaiu-
shchei promyshlennosti Rossii i ikh
antirabochaia politika v gody pervoi mirovoi
voiny' (Moscow State University candidate's
dissertation, 1964, deposited in GBIL) pp.
121-2, interprets this proposal as an attempt
by the Petrograd 'monopolists' to take over
the TsVPK.

73 This is the interpretation of J.D. White,
'Moscow, Petersburg and the Russian Indus-
trialists', p. 419. The other delegate was
Litvinov-Falinskii.

74 *Gornozavodskoe delo*, no. 29, 22 July 1915, pp.
11481-2 supported Avdakov's proposal; *Utro
Rossii*, no. 172, 24 June 1915, p. 3 opposed
it.

75 'He [Guchkov] always used his business ties
and contracts to further his political aims,
not *vice versa*' - L. Menashe, 'Alexander
Guchkov and the Origins of the Octrobrist
Party' (New York University Ph.D. dissertation, 1966) p. 80. Guchkov's business ties
included shares in, and board membership on,
the 'Russia' Insurance Company, the 'Skoro-
khod' Shoe Company of Petrograd, and the *Novoe
Vremia* publishing firm. (ibid., p. 23).

76 V.I. Lenin, *Polnoe sobranie sochinenii*, 55
vols (Moscow, 1958-65) vol. XXXI, p. 302. Kit
Kitich, or Tit Titich, was a petty tyrant who
figures in 'Carrying the Can for Others', a
play by A.N. Ostrovskii; L. Trotsky, *The
History of the Russian Revolution*, trans. M.

Eastman, 3 vols (London, 1967) vol. I, p. 82.
77 Menashe, 'Alexander Guchkov and the Origins of
    the Octobrist Party', p. 67. There is a long
    autobiographical letter from Guchkov recount-
    ing his battlefield experiences in Baron
    Aleksandr Feliksovich Meyendorff Papers,
    Valtionarkisto (Finnish State Archives),
    Helsinki, Kansio 2, correspondence with A.I.
    Guchkov, 8 Feb. 1931.
78 Guchkov's speeches were republished as *A.I.
    Guchkov v tret'ei Gosudarstvennoi dume, 1907-
    1912 gg.* (St Petersburg, 1912). Additional
    speeches are included in a collection
    published while Guchkov was War Minister in
    the Provisional Government. See A.I. Guchkov,
    *Rechi po voprosam Gosudarstvennoi oborony i ob
    obshchei politike, 1908-17* (Petrograd, 1917).
79 Gos. Duma, III Sozyv, *Sten. otchety*, ses. 1,
    zas. 74 (27 May 1908) col. 1596.
80 A.I. Guchkov, *Doklad po zakonoproektu ob
    otpuske na 1912 goda sredstv na popolnenie
    zapasov i materialov (sekretno)* (St Peters-
    burg, 1912).
81 Gos. Duma, III Sozyv, *Sten. otchety*, ses. 5,
    zas. 126 (7 May 1912) col. 735.
82 'Perepiska V.A. Sukhomlinova s N.N. Ianush-
    kevichem', *KA*, no. 2 (1922) p. 166, letter of
    10 Jan. 1915; 'Iz dnevnika Sukhomlinova', *Dela
    i dni*, no. 1 (1920) p. 230. B. Pares, *My
    Russian Memoirs* (London, 1931) pp. 328-9 is
    more charitable to Guchkov.
83 FO 371 2454 N116039/105582: '"Report on
    (unpublished) speech on 26 June by Mr A.
    Guchkov on Russian munitions" by Bernard
    Pares, 17/30 June 1915'.
84 *Trudy s"ezda predstavitelei VPK-ov*, p. 109.
85 Ibid., pp. 111-14.
86 Ibid., pp. 118, 129. *Promyshlennaia Rossiia*,
    which claimed to be the voice of small- and
    medium-scale industry, interpreted the con-
    gress as a struggle against the 'monopolists'.
    'Monopolists in the industrial sector', it
    asserted, 'with rare exceptions such as P.P.
    Riabushinskii, remain monopolists in public
    activity....Counting on the (committees')
    importance for future economic growth, the
    gentlemen from Liteiny want to subordinate
    them to their control' ('God voiny i promy-

shlennost', '*Promyshlennaia Rossiia*, no. 14-15, 26 July 1915, pp. 3-4).

87   *Trudy s"ezda predstavitelei VPK-ov*, p. 125.

88   This dichotomy between 'old foreign-based capital' and new domestic capital as reflected in the dispute between the Association and the VPKs was first discussed in a rarely cited but extremely interesting article by V. Reikhardt, 'K probleme monopolisticheskogo kapitalizma v Rossii - evoliutsiia uchastiia inostrannogo kapitala v russkom narodnom khoziaistve za gody imperialisticheskoi voiny, 1914-1917', *Problemy marksizma*, no. 5-6 (1931) pp. 193-222.

89   *Trudy s"ezda predstavitelei VPK-ov*, pp. 143-5.

90   'Mobilizatsiia promyshlennosti', *PT*, no. 8 (183), 8 Aug. 1915, p. 133. Among the participants in this section were Fon-Ditmar and Eiler representing coal interests, Maidel and M.F. Norpe from the Association, and A.O. Gukasov of Baku.

91   *PT*, no. 8 (183), 8 Aug. 1915, p. 133.

92   Ibid., p. 129.

93   *Trudy s"ezda predstavitelei VPK-ov*, p. 110.

94   *PT*, no. 8 (183), 8 Aug. 1915, pp. 128-9.

NOTES TO CHAPTER 4:   THE HIGH POLITICS OF DEFENCE

1   T.D. Krupina, 'Politicheskii krizis', p. 65. There are a number of discrepancies between the table presented by Krupina and the account in S.A. Somov, 'O "maiskom" Osobom soveshchanii', *Istoriia SSSR*, no. 3 (1973) p. 119. Krupina is more reliable.

2   Somov, 'O "maiskom" Osobom soveshchanii', p. 119. Konovalov had been chosen by the Moscow VPK on 30 June with Tret'iakov as his assistant.

3   *ZhOSO*, part 1, no. 2, 18 May 1915, p. 12; no. 4, 27 May 1915, p. 24; no. 6, 6 June 1915, p. 47; no. 7, 13 June 1915, p. 61.

4   Ibid., no. 19, 31 July 1915, pp. 164-6.

5   Somov, 'O "maiskom" Osobom soveshchanii', p. 121; *ZhOSO*, part 1, no. 12, 4 July 1915, p. 104.

6   *Russkie Vedomosti*, no. 191, 24 Aug. 1915, p. 4; *ZhOSO*, part 1, no. 21, 8 Aug. 1915, p. 182

and no. 23, 15 Aug. 1915, p. 194.
7   Somov, 'O "maiskom" Osobom soveshchanii', p. 122.
8   'Spisok zakazov poluchennykh mekhanicheskim otdelom TsVPKa' (22 Aug. 1915), TsGVIA, f. 369, op. 4, d. 20, 1. 53. The TsVPK did everything but sign the contracts which were concluded directly by the GAU and the engineering unit with enterprises.
9   *Utro Rossii*, no. 167, 19 June 1915, p. 1.
10 *Birzhevye Vedomosti*, no. 14951, 8 July 1915, p. 2. S.V. Voronkova, 'Materialy Osobogo soveshchaniia po oborone gosudarstva' (Moscow State University candidate's dissertation, 1970, deposited in GBIL), p. 57 claims that the plan arose at the end of June or early July.
11 Voronkova, 'Materialy', p. 60. The final version is in 'Proekt', TsGIA, f. 1276, op. 11, d. 888, 11. 11-21.
12 *Utro Rossii*, no. 197, 19 July 1915, p. 1.
13 'Doklad fraktsii Progressistov', GBIL, f. 260, kart. 4, d. 10, 11. 7-8; *Rech'*, no. 159, 12 June 1915, p. 3.
14. *Gornozavodskoe delo*, no. 30, 31 July 1915, pp. 11520-1.
15 Razumovskaia, 'Tsentral'nyi voenno-promyshlennyi komitet', p. 47, asserts that the Committee of State Defence was to be an expanded TsVPK. Diakin, *Russkaia burzhuaziia i tsarizm*, pp. 87-8, points out, however, that it was precisely because the VPKs were to have only a 'small voice' that their representatives balked at the project.
16 TsGIA, f. 1276, op. 11, d. 888, 1. 176.
17 'Kantseliariia gosudarstvennoi dumy predsedateliu Soveta ministrov' (28 July 1915), TsGIA, f. 1276, op. 11, d. 888, 1. 43.
18 *Trudy s"ezda predstavitelei VPK-ov*, pp. 68-70. Much of this speech was censored from the published stenographic report, the only one available.
19 Ibid., pp. 34-5.
20 *Russkie Vedomosti*, no. 171, 25 July 1915, p. 4.
21 *Trudy s"ezda predstavitelei VPK-ov*, pp. 32, 77.
22 Ibid., pp. 57, 97. The powers described in the

resolution were so broad the Margulies could
only wonder what would become of the Minister
of War. For the resolution see *PT*, no. 6
(183), 8 Aug. 1915, pp. 129-30.

23   *Russkoe Slovo*, no. 171, 25 July 1915, p. 4
cites the meeting as a session of the Moscow
Exchange Committee, the original sponsor of
the Moscow VPK.

24   'Iz zapisnoi knizhki arkhivista', *KA*, no. 59
(1933) p. 147.

25   *Birzhevye Vedomosti*, no. 14907, 16 June 1915,
p. 2; ibid., no. 14945, 20 July 1915, p. 2;
Russkoe Slovo, no. 170, 24 July 1915, p. 5.
Other names mentioned for this post were
Shingarev and Litvinov-Falinskii (see *Utro
Rossii*, no. 204, 26 July 1915, p. 1).

26   'Protokoly zasedaniia ts. k. partii za avgusta
1915 g.', GBIL, f. 225, papka 5, l. 8 obv.

27   Countess Panina Collection, Columbia University Archive of Russian and East European
History and Culture, papka 1, folder 6, letter
40. This antipathy dated back to Guchkov's
famous speech delivered to the Third Duma's
Commission on State Defence in May 1908.
Guchkov attacked the interference of the Grand
Dukes in military affairs and cited this as a
primary reason for the administrative confusion during the Japanese War. Later, during
the First World War, the Empress was to muse
'Could not a strong railway accident be
arranged in which he [Guchkov] alone would
suffer?' (see *Pis'ma Imperatritsy Aleksandry
Fedorovny k Imperatoru Nikolaiu II*, ed. V.D.
Nabokov, 2 vols (Berlin, 1922) vol. I, p.
626).

28   A.A. Polivanov, *Iz dnevnikov i vospominanii*,
p. 207.

29   'Polozheniia', *Izv. MVPKa*, no. 2 (Sept. 1915)
pp. 21-30.

30   The Commissions, either separately or jointly,
it is not clear which, met three times after
the session on 24 July. For brief announce-
ments, see *Russkie Vedomosti*, no. 172, 26 July
1915, p. 2; 'Den', no. 205, 28 July 1915, p.
3; *Utro Rossii*, no. 209, 31 July 1915, p. 2.
These were closed sessions and very little
information was provided in the press. It is
likely that the decision was taken at the

session on 30 July.

31 Gos. Duma, IV Sozyv: *Sten. otchety*, ses. 4, zas. 4 (1 Aug. 1915) cols 300-3, 324.

32 Gosudarstvennyi Sovet: *Stenograficheskie otchety*, sessiia 11, zasedanie 5 (17 Aug. 1915) cols 109-12, 115; *Novoe Vremia*, no. 14166, 18 Aug. 1915, p. 4.

33 *Izv. TsVPKa*, no. 1, 24 Aug. 1915, p. 1.

34 'Lukomskii voennomu ministru' (2 Aug. 1915), TsGVIA, f. 369, op. 1, d. 72, l. 43.

35 'Dokladnaia zapiska po GUGSh, otdel po ustroistvu i sluzhbe voisk' (3 Aug. 1915), TsGVIA, f. 369, op. 1, d. 72, l. 48.

36 A.N. Iakhontov, 'Tiazhelye dni: sekretnye zasedaniia Soveta ministrov 16 iiulia-2 sentiabria 1915 g.', *ARR*, vol. XVIII (Berlin, 1926) p. 36.

37 The version of 30 July drawn up by the committees and the final version of 27 August are juxtaposed in TsGVIA, f. 369, op. 1, d. 72, ll. 40-3, 66-9.

38 Iakhontov, 'Tiazhelye dni', p. 59. These remarks were made at the session of the Council of Ministers on 9 August.

39 See *KA*, no. 50-1 (1932); no. 52 (1932); no. 56 (1933); no. 59 (1933).

40 Cf. I.F. Gindin, 'Problemy istorii fevral'skoi revoliutsii i ee sotsial'no-ekonomicheskykh predposilok', *Istoriia SSSR*, no. 4 (1967) pp. 40-69; and V.S. Diakin, 'Chto takoe Progressivnyi blok?', *Voprosy istorii*, no. 1 (1970) pp. 200-4.

41 P.N. Miliukov, *Rossiia na perelome*, 2 vols (Paris, 1927) vol. I, pp. 12-13; Laverychev, *Po tu storonu barrikad*, p. 121.

42 *Burzhuaziia nakanune fevral'skoi revoliutsii*, ed. B.B. Grave (Moscow and Leningrad, 1927) pp. 62, 65. 'Kadety v dni Galitsiiskogo razgroma 1915 g.', ed. N. Lapin, *KA*, no. 59 (1933) p. 119.

43 Ibid., p. 127. This statement was made by Nekrasov.

44 'Progressivnyi blok v 1915-1917 gg.', ed. N. Lapin, *KA*, no. 50-1 (1932) p. 150.

45 *Utro Rossii*, no. 222, 13 Aug. 1915, p. 2.

46 *Burzhuaziia nakanune*, p. 21; resolution quoted in Pares, *Fall of the Russian Monarchy*, p. 263.

47 Laverychev, *Po tu storonu barrikad*, pp. 119-
    20; *Rech'*, no. 229, 20 Aug. 1915, p. 3; *Utro
    Rossii*, no. 226, 17 Aug. 1915, p. 3.
48 *Burzhuaziia nakanune*, p. 25.
49 Ibid., p. 24.
50 'TsVPK biuro, tret'ia sessiia' (24 Aug. 1915),
    TsGVIA, f. 369, op. 1, d. 205, l. 9.
51 FO 371 2454 N125874: 'Buchanan to Grey, 23
    Aug./5 Sept. 1915'. It is unclear which com-
    mittees Buchanan had in mind. He refers to the
    'war-work committees' which could mean the
    VPKs as well as the Special Councils. For
    Shingarev's memorandum, see Semennikov,
    Monarkhiia pered krusheniem, pp. 267-75.
52 Krupina, 'Politicheskii krizis', p. 75, claims
    that the Tsar's display of humility of 22
    August was part of his ruse.
53 *Burzhuaziia nakanune*, pp. 29-31; see also V.I.
    Startsev, *Russkaia burzhuaziia i samoderzhavie
    v 1905-1917 gg. (Bor'ba vokrug 'otvetstvennogo
    ministerstva' i 'Pravitel'stva doveriia')*
    (Leningrad, 1977) pp. 137-8.
54 *Burzhuaziia nakanune*, pp. 38, 41; A.Ia. Grunt,
    'Progressivnyi blok', *Voprosy istorii*, no. 3-4
    (1945) p. 114.
55 M. Balabanov, *Ot 1905 k 1917: Massovoe rab-
    ochee dvizhenie* (Moscow and Leningrad, 1927)
    p. 409.
56 Diakin, *Russkaia burzhuaziia i tsarizm*, p.
    120; *Burzhuaziia nakanune*, p. 41.
57 S.L. Gisin, 'Vserossiiskii zemskii soiuz:
    Politicheskaia evoliutsiia s iiulia 1914 po
    fevral' 1917' (candidate's dissertation, 1946,
    deposited in GBIL) p. 163; *Rech'*, no. 248, 9
    Sept. 1915, p. 4.
58 *Burzhuaziia nakanune*, p. 50.
59 *Utro Rossii*, no. 266, 8 Sept. 1915, p. 3. On
    the delegation and related attempts to remon-
    strate with the Tsar, see Startsev, *Russkaia
    burzhuaziia i samoderzhavie*, pp. 179-82.

NOTES TO CHAPTER 5: THE WAR-INDUSTRIES COMMITTEES
AND MILITARY SUPPLIES

1   Quoted in M. Ia. Gefter, 'Tsarizm i monopol-
    isticheskii kapital v metallurgii Iuga
    Rossii', *IZ*, vol. XLIII (1953) p. 89. The

official was A.A. Vol'skii of the Advisory Office of Iron Mill Owners and the occasion was a special meeting called by the government in May 1908 to assess the condition of the metallurgical and metalworks industries. See also R.A. Roosa, 'The Association of Industry and Trade, 1906-1914' (Columbia University Ph.D. dissertation, 1967) pp. 413, 515.

2 For the question of state socialism in pre-revolutionary Russia, see Roosa, 'Russian Industrialists and "State Socialism"', pp. 395-417. Lenin had some interesting remarks on state orders, particularly military contracts: 'Armaments is considered a national endeavour, a patriotic duty...but shipbuilding, gunpowder, dynamite and arms factories are above all international enterprises in which the capital of various countries gladly swindles and fleeces "the public" building ships or cannons identically for England against Italy, for Italy against England' (*Sochineniia*, 4th edn, vol. XIX, pp. 83-4).

3 Roosa, 'The Association of Industry and Trade, 1906-1914', p. 406.

4 'Spisok soveshchanii soveta s"ezdov s ukazaniem rassmatrivaemykh voprosov i predstavitelei zasedaniia, 1909-1910', TsGIA, f. 32, op. 2, d. 111, ll. 1-10.

5 'Ispravlennoe polozhenie' (11 May 1909), TsGIA, f. 32, op. 1, d. 398, ll. 130-53; *PT*, no. 23, 1 Dec. 1909, pp. 583-5.

6 'Khronika deiatel'nosti Pravitel'stva', *PT*, no. 19 (163), 1 Oct. 1914, p. 336.

7 *Russkie Vedomosti*, no. 137, 16 June 1915, p. 3.

8 Ia.M. Bukshpan, 'Voennyi komitet nemetskoi promyshlennosti', *PT*, no. 11 (179), 1 June 1915, pp. 582-5. Bukshpan's article was not the only one to call the attention of Russian industrialists to the German arrangement. Cf., for instance, *Rech'*, no. 148, 31 May 1915, p. 1. The coincidence in dates between the two articles and the founding of the VPKs should be noted.

9 Feldman, *Army, Industry, and Labor in Germany*, p. 45.

10 At the VPKs' third congress in May 1917, Guchkov went so far as to assert that 'we

never took it into our heads to be the sole supplier and contractor of the Ministry of War' (see *Izv. TsVPKa*, no. 230, 25 May 1917, p. 2).

11 'Polozhenie predstavitelei Novonikolaevskogo VPKa (Tomsk)' (15 Feb. 1916), TsGVIA, f. 13251, op. 1, d. 8, 1. 95; *Izv. MVPKa*, no. 11, Dec. 1915, p. 48; and Ia.A. Galiashkin, 'O finansovom polozhenii VPKov', *Izv. MVPKa*, no. 34, Apr. 1917, p. 29.

12 For the former see Tsentral'noe statisticheskoe upravlenie, *Rossiia v mirovoi voine* (Moscow, 1925) Table 49; for the latter see I.V. Maevskii, *Ekonomika russkoi promyshlennosti v usloviiakh pervoi mirovoi voiny* (Moscow, 1957) p. 93.

13 For February 1917 figures see *Otchety otdelov TsVPKa k tret'emu Vserossiiskomu s"ezdu VPK-ov v Moskve* (Petrograd, 1917) pp. 133, 201; for October estimate see *Izv. MVPKa*, no. 48-9, Dec. 1917, p. 12; and for final figure based on a report to the fourth congress, see I.A. Gorbachev, *Khoziaistvo i finansy voennopromyshlennykh komitetov* (Moscow, 1919) p. 23. A grant of 150 million rubles made by the Provisional Government in October 1917 accounts for much of the difference between the last two figures.

14 'Sravnitel'naia vedomost' stoimosti zakazov voennogo vedomstva s nachala voiny do nachala deiatel'nosti VPK-ov i s etogo sroka po 1 maia 1916 g.', TsGVIA, f. 369, op. 4, d. 242, 1. 86.

15 Russian enterprises obtained orders valued at 2,495,265,900 rubles in 1916 according to *Rossiia v mirovoi voine*, Table 47, p. 56.

16 'Chetvertoe zasedanie TsVPKa' (31 Aug. 1915), TsGVIA, f. 369, op. 1, d. 205, 1. 12. The plenipotentiary was I. Lavrov, chairman of the Irkutsk district VPK.

17 *Doklad schetnogo otdela, 26 fevralia 1916 g.* (Petrograd, 1916) p. 13; *Soobschcheniia predstavitelei 34 voenno-promyshlennykh komitetov, 7-31 dekabria, 1915 g.* (Petrograd, 1916) pp. 91, 104.

18 *Trudy vtorogo s"ezda*, p. 639; *Doklad schetnogo otdela na 26 sentiabria 1916 g.* (Petrograd, 1916) p. 21.

19 *Statisticheskaia   razrabotka   dannykh   o deiatel'nosti glavnogo po snabzheniiu armii komiteta Vserossiiskogo Zemskogo Soiuza i Vserossiiskogo Soiuza Gorodov i voenno-promyshlennykh komitetov po vypolneniiu plan-ovykh zakazov voennogo vedomstva pervoi ocheredi* (Moscow, 1917) p. 19.

20 *Otchety otdelov TsVPKa k tret'emu s"ezdu*, p. 117.

21 *Soobshcheniia 34 VPK-ov*, pp. 48, 57.

22 Razumovskaia,   'Tsentral'nyi   voenno-promy-shlennyi komitet', p. 125. The corresponding figures by August 1916 were 177.14 million and 26.53 million according to *Finansovyi otchet s nachala deiatel'nosti MVPKa*, p. 1.

23 *Deiatel'nost' Moskovskogo oblastnogo VPKa i ego otdelov na 1/IV 1917 g.* (Moscow, 1917) p. 40. Figures as of June 1917 are presented in *Izv. MVPKa*, no. 38, July 1917, p. 4.

24 'Istoricheskaia   spravka   o   deiatel'nosti Moskovskogo VPKa', *Izv. MVPKa*, no. 33–4, June 1916, p. 68; 'Doklad upravliaiushchego delami komiteta A.I. Sinelobova k plenarnomu zasedan-iiu komiteta 19/II (4/III) 1918: Deiatel'nost' Moskovskogo VPKa', *Izv. TsVPKa*, no. 293, 18 Apr. 1918, p. 2; and *Izv. MVPKa*, no. 47, Nov. 1917, p. 8.

25 *Deiatel'nost' oblastnykh i mestnykh voenno-promyshlennykh komitetov na 10 fevralia 1916 goda* (Petrograd, 1916) vol. I, pp. 34, 70.

26 *Izvestiia glavnogo komiteta po snabzheniiu armii*, no. 15–6, 15 Apr. 1916, pp. 74–6.

27 *Istoriia organizatsii S.N. Vankova*, pp. 166–8, 170–1, 180, 209.

28 *Izv. MVPKa*, no. 17–18, March 1916, p. 135. In a letter to the Moscow VPK, General N.I. Bogatko, chief of the GIU, claimed that orders from individual units for clothing 'would be excessive' (see *Russkoe Slovo*, no. 78, 5 Apr. 1916, p. 5).

29 'Doklad praporshchika Pisareva TsVPKu' (16 May 1916), TsGAOR, f. 555, op. 1, d. 289, ll. 1–4.

30 'Zhurnal(y) zasedanii biuro TsVPKa', nos 267, 322, 405 (1 July 1916, 16 Sept. 1916 and 10 Jan. 1917), TsGVIA, f. 13251, op. 1, d. 46, l. 1 obv.; d. 47, ll. 16 obv., 27; 'Soveshchanie predstavitelei oblastnykh VPK-ov', *PT*, no. 40 (233), 8 Oct. 1916, p. 269.

31 *Statisticheskaia razrabotka*, p. 10.
32 'Predpriiatiia poluchaiushchie zakazy ot mekhanicheskogo otdela', TsGVIA, f. 369, op. 4, d. 20, l. 64 obv. Enterprises receiving orders directly from the section numbered 144. There were 297 which received them through provincial committees. The list of 55 items is in 'Vedomost' zakazov na snariady, granaty, bombomety, minomety, i pr. na 1 fev. 1916 g. (mekhanicheskii otdel)', TsGVIA, f. 369, op. 21, d. 49, l. 50.
33 *Rossiia v mirovoi voine*, Table 55, p. 61; Table 49, p. 58.
34 Manikovskii, *Boevoe snabzhenie*, vol. I, Table 19, pp. 250-2.
35 G.K. Miftiev, 'Artilleriiskaia promyshlennost' Rossii v period pervoi mirovoi voiny' (candidate's dissertation, 1953, deposited in GBIL) p. 185. The Central Statistical Administration did not include mortars in its list of articles produced under the TsVPK's auspices.
36 'Mobilizatsiia i ob"edinenie', *Promyshlennaia Rossiia*, no. 10-11, 28 June 1915, pp. 1-2.
37 *Organizatsiia raspredeleniia voennykh zakazov po raionam Imperii* (Petrograd, 1915) p. 4.
38 'Ot predsedatelia Kievskogo Zemstva Osobomu soveshchaniiu' (8 July 1915), TsGVIA, f. 369, op. 1, d. 72, l. 32. Bobrinskii derived his wealth from sugar manufacturing. He was the leader of the Nationalist Party and in 1916 was appointed Minister of Agriculture after serving as Assistant Minister of Internal Affairs for several months.
39 Zagorsky, *State Control of Industry*, p. 103.
40 'Zapiska predsedatelia TsVPKa Lukomskomu' (4 Jan. 1916), TsGVIA, f. 369, op. 4, d. 117, ll. 96-8; 'Zapiska o zatrudneniiakh v dostavlenii materialov s Putilova, Obukhovskogo, i Sormova' (11 Feb. 1916), TsGAOR, f. 555, op. 1, d. 359, ll. 1-2 obv.
41 Pogrebinskii, for example, contends that 'the VPKs did not play a serious role in the business of supplying the army'. See his 'K istorii soiuzov zemstv i gorodov v gody imperialisticheskoi voiny', *IZ*, vol. XII (1941) p. 47. Sidorov, 'Bor'ba s krizisom vooruzheniia', p. 47, characterizes the committees as 'an

intermediary between the state and small-scale industry'. See also Stone, *The Eastern Front*, pp. 201-5 which relies heavily on, and agrees with, Pogrebinskii and Sidorov.

42 Razumovskaia, 'Tsentral'nyi voenno-promyshlennyi komitet', p. 114, quoting TsGVIA, f. 369, op. 16, d. 439, ll. 15, 17, 27, 29.

43 TsGVIA, f. 369, op. 21, d. 49, ll. 61-6.

44 Korelin, 'Monopolii v metal. prom.', p. 178.

45 *Aktsionerno-paevye obshchestva v Rossii* (Moscow, 1913) pp. 100, 219, 346, 456.

46 TsGAOR, f. 102 (1916), d. 343 (2), ll. 104-5. The letter, dated 17 May 1916, was sent from the Crimea where the elder Riabushinskii was recuperating from a respiratory disease.

47 Razumovskaia, 'Tsentral'nyi voenno-promyshlennyi komitet', p. 125, quoting TsGVIA, f. 369, op. 1, d. 568, l. 211.

48 *Spisok fabrik i zavodov g. Moskvy i Moskovskoi gubernii sostavlen fabrichnymi inspektorami Moskovskoi gubernii po dannym 1916 goda* (Moscow, 1916). The list does not include state enterprises or those of public institutions.

49 I.I. Olovianishnikov, presumably the brother of P.I., was chairman of the Moscow VPK's evacuation and relocation section.

50 *Izv. MVPKa*, no. 27-30, Aug.-Sept. 1916, p. 310. Fabergé, world-renowned for the *'bonbons'* it produced for the royal family, also accepted orders from the Vankov Organization.

51 'Upolnomochii predsedatelia Osobogo soveshchaniia o zakazakh Kievskogo VPKa' (21 May 1916), TsGVIA, f. 369, op. 21, l. 295.

52 See p. 80. This stipulation did not, however, rule out profit through the reduction of inventory and labour costs.

53 *Spisok sobstvennykh predpriiatii voennopromyshlennykh komitetov* (Petrograd, 1917) pp. 28-9; 'Inzhiner tekhnolog N.V. Monakhov, v Biuro TsVPKa o deiatel'nosti Samarskogo VPKa' (4 July 1916), TsGAOR, f. 555, op. 1, d. 294, ll. 1-6.

54 *Izv. MVPKa*, no. 15-16, Feb. 1916, p. 133. By 1 June 1916 the committee received 1.4 million rubles as advances, a sum which would have just covered the purchase of the factories. See *Izv. MVPKa*, no. 23-4, June 1916, p. 216.

55 *Spisok sobstvennykh predpriiatii VPK-ov*, pp. 12–15; 'TsVPK Osobomu soveshchaniiu po voprosu ob otpuske 13 millionov rublei v vide oborotnykh sredstv dlia Moskovskogo VPKa' (13 June 1917), TsGVIA, f. 369, op. 1, d. 402, l. 17.

56 *Spisok sobstvennykh predpriiatii VPK-ov*, pp. ix, x.

57 Pogrebinskii, 'Voenno-promyshlennye komitety', p. 175.

58 *Spisok sobstvennykh predpriiatii VPK-ov*, pp. v, 10–11, 14–15, 28–9, 38–9, 46–7. The Baku plant was one of three established in 1916 to produce toluene as an oil extract. A.O. Gukasov financed the operation. For details see Ipat'ev, *Zhizn' odnogo khimika*, vol. I, pp. 474–5.

59 *Spisok sobstvennykh predpriiatii VPK-ov*, pp. 44–5. '22-oe zasedanie TsVPKa' (13 June 1916), TsGVIA, f. 13251, op. 40, d. 30, l. 153. In 1917 'Protivogaz' was also placed under the TsVPK's administration.

60 'Spravka o kolichestve respiratorov Zelinskogo-Kummanta', TsGVIA, f. 369, op. 1, d. 206, l. 416.

61 Ipat'ev, *Zhizn' odnogo khimika*, vol. I, pp. 472–3.

62 'Spisok lits, rabotaiushchikhsia na otdel izobretenii TsVPKa', TsGVIA, f. 13251, op. 1, d. 7, l. 67.

63 *Izv. MVPKa*, no. 6–7, Oct. 1915, p. 80; no. 9–10, Nov. 1915, p. 6; no. 31–2, Oct.–Nov. 1916, p. 44.

64 Ibid., no. 47, Nov. 1917, p. 8.

65 Ibid., no. 8, Nov. 1915, pp. 22–6.

66 Ibid., no. 27–30, Aug.–Sept. 1916, pp. 219–20; see also, *Proizvoditel'nye sily Rossii*, no. 1, Nov. 1916, pp. 26–9; *Inzhener'* (Kiev), no. 2, Feb. 1917, pp. 61–3.

67 'Dokladnaia zapiska ot TsVPKa po vyiasneniiu prichin nesobliudeniia zavodami srokov postavki snariadov', TsGVIA, f. 369, op. 1, d. 72, l. 34.

68 'Doklad schetnogo otdela' (4 July 1916), TsGVIA, f. 369, op. 4, d. 242, l. 29.

69 'Zhurnal shestogo zasedaniia TsVPKa' (21 Sept. 1915), TsGVIA, f. 369, op. 1, d. 205, ll. 18, 20.

70 *Den'*, no. 293, 24 Oct. 1915, p. 2.

71 Quoted in FO 371 2745 N5724: 'Lockhart to Buchanan', 2/15 Dec. 1915. See also *Trudy Vserossiiskogo monarkhicheskogo soveshchaniia v g. Nizhnem Novgorode upolnomochennykh pravykh organizatsii s 26 po 29 noiabria 1915 g.* (Petrograd, 1916) prilozhenie 8, p. 24, where explicit reference is made to Manikovskii's interview.

72 'Bespechnost'', *Den'*, no. 317, 17 Nov. 1915, p. 2.

73 *Rech'*, no. 334, 4 Dec. 1915, p. 5. The publicist was Professor N.N. Savvin, later Assistant Minister of Trade and Industry in the Provisional Government; for the endorsement see 'Pri svet glasnosti', *Novoe Vremia*, no. 14276, 6 Dec. 1915, p. 5.

74 'Vedomost' o sostoianii zakazov priniatykh k ispolneniiu TsVPK-om sroki sdachy koikh naznacheny TsVPK-om do 1 dek. 1915 g.', TsGVIA, f. 369, op. 4, d. 20, ll. 5 obv.-6, 69 obv.-70. Of the sixteen articles listed only landmine containers had been delivered on time.

75 *Istoriia organizatsii S.N. Vankova*, pp. 152, 157-60.

76 Figures for January in 'Vedomost' o kolichestve predmetov priniatykh ot postavshchikov TsVPKa na 22 ianv. 1916 g. dlia GAU', TsGVIA, f. 369, op. 21, d. 49, l. 18; for February, 'Svedenie o kolichestve vypolnenykh VPK-om postavok po 22 fev. 1916', ibid., op. 4, d. 117, l. 206; for March, 'Vedomost' predmetov izgotovlennykh postavshchikami i priniatykh komitetom dlia GAU na 22 marta', ibid., op. 21, d. 49, l. 204.

77 *Deiatel'nost' oblastnykh i mestnykh VPK-ov na 10 fev. 1916 goda*, vol. I, pp. 37-9; vol. II, p. 51; vol. III, p. 167.

78 'Sibirskoe zavodskoe soveshchanie' (20 Feb. 1916), TsGVIA, f. 369, op. 4, d. 117, l. 267.

79 'Svedenie o khode priemki i sdachy izgotovlennykh predmetov', TsGAOR, f. 102 (1916), d. 338, l. 44 obv; *Izv. MVPKa*, no. 21-2, May 1916, p. 174; ibid., no. 25-6, July 1916, p. 51; *Izv. TsVPKa*, no. 100, 26 May 1916, p. 4; 'Kavkazskoe zavodskoe soveshchanie' (15 Apr. 1916), TsGVIA, f. 369, op. 4, d. 117, l. 451.

80 'Zhurnal 19-ogo zasedaniia TsVPKa' (18 Mar. 1916), TsGVIA, f. 369, op. 1, d. 205, l. 255; Sirinov, 'Pobol'she dobrosovestnosti', *Izv. TsVPKa*, no. 95, 14 May 1916, p. 1.

81 'Vedomost' izgotovlennykh i priniatykh predmetov po zakazam TsVPKa dlia GAU' (8 July 1916), TsGVIA, f. 369, op. 21, d. 49, l. 346.

82 *Izv. TsVPKa*, no. 183, 20 Dec. 1916, p. 2.

83 Sidorov, *Ekonomicheskoe polozhenie*, p. 203. According to Gorbachev, *Khoziaistvo i finansy VPK-ov*, pp. 23-4, by January 1918 the total value of orders fulfilled by the VPKs was 290 million rubles or 48 per cent of the amount distributed by the TsVPK. The Moscow committee fulfilled orders valued at 119 million rubles, out of a total of 229 million rubles in orders distributed.

84 *Izv. glav. kom. po snab. armii*, no. 15-16, 15 Apr. 1916, p. 29.

85 *Izv. MVPKa*, no. 27-30, Aug.-Sept. 1916, p. 125.

86 'Dokladnaia zapiska o sdache T/D. Pitoeva zapasa na 3 milliona ruchnykh granatov (1914)' (5 May 1916), TsGAOR, f. 555, op. 1, d. 362, ll. 1-2; TsGVIA, f. 369, op. 4, d. 117, ll. 485-6.

87 *Rech'*, no. 341, 11 Dec. 1915, p. 4; 'Soveshchanie predstavitelei oblastnykh VPK-ov, 23 maia', *Izv. TsVPKa*, no. 100, 26 May 1916, p. 3.

88 *Izv. MVPKa*, no. 15-16, Feb. 1916, p. 140.

89 'Osobyi zhurnal Soveta ministrov' (3 Nov. 1915), TsGVIA, f. 369, op. 1, d. 205, l. 47.

90 Cf. Manikovskii, *Boevoe snabzhenie*, vol. I, p. 253; Maevskii, *Ekonomika russkoi promyshlennosti*, pp. 290-1; S.P. Borisov, *Bor'ba bol'shevikov protiv voenno-promyshlennykh komitetov* (Moscow, 1948) p. 12.

91 Tarnovskii, *Formirovanie*, p. 159. For Prodamet's upward revision of prices, see *Monopolii v metallurgicheskoi promyshlennosti Rossii 1900-1917, Dokumenty i materialy* (Moscow and Leningrad, 1963) pp. 158-9, 160-1, 165-77.

92 *Statisticheskaia razrabotka*, p. 52.

93 'Vedomost' tsen predmetov zakazennykh mekhanicheskim otdelom', TsGVIA, f. 369, op. 4, d. 20, l. 65; 'Vedomost' na predmety GAU i

obshchestvennykh organizatsii', ibid., op. 21, d. 108, l. 33.

94 'Veshchevoi otdel', TsGVIA, f. 369, op. 4, d. 20, l. 33; 'Vedomost' na predmety GIU i obshchestvennykh organizatsii', ibid., op. 21, d. 108, l. 21.

95 'TsVPK Shuvaevu' (8 Nov. 1916), TsGVIA, f. 369, op. 1, d. 279, l. 52.

96 '5 fev. 1916' (5 Feb. 1916), ibid., d. 120, l. 45. The authorship of this memorandum is in some doubt. Tarnovskii (*Formirovanie*, p. 174) suggests General E.K. Hermonius, chairman of the Russian Governmental Committee in London, but the copy in TsGVIA which I saw was signed by 'Markov'.

97 'Doklad Pereverzeva o nedorazumenii v Nizhegorodskom VPKe, Pavlovskom raione', TsGAOR, f. 102 (1916), d. 343 (2), l. 260. See also *Den'*, no. 248, 9 Sept. 1915, p. 3.

98 *Izv. MVPKa*, no. 19-20, Apr. 1916, p. 211.

99 'Doklad o deiatel'nosti Arkhangel'skogo oblastnogo VPKa, L. Tlushchikovskii' (21-9 July 1916), TsGAOR, f. 555, op. 1, d. 292, ll. 1-8 obv.

100 TsGVIA, f. 369, op. 4, d. 117, l. 481; 'Spravka o zakaze na 48-lin i 6-dm. chugunye snariady (dlia Bar. G.Kh. Maidela)', ibid., op. 1, d. 206, l. 332.

101 *Trudy vtorogo s"ezda*, p. 93. In December 1915 the GIU granted the committees the right to appoint their own inspectors. See *Izv. MVPKa*, no. 11, Dec. 1915, p. 40.

102 'Protokoly mekhanicheskogo otdela Moskovskogo VPKa' (19 Feb. 1916), TsGVIA, f. 13251, op. 1, d. 8, ll. 99-104 obv.

103 *Izv. TsVPKa*, no. 175, 1 Dec. 1916, p. 1. In June 1916 a report reached the police of a complaint filed by workers of the Olovianishnikov factory against a GAU inspector. The inspector had refused a consignment of 30,000 shells on insufficient grounds and was dismissed. The report is in '"Pavlov" Klimovichu' (30 June 1916), TsGAOR, f. 102 (1916), d. 343 (2), l. 255. For a report of another incident see *Izv. TsVPKa*, no. 95, 14 May 1916, p. 1.

104 *Sobranie uzakonenii i rasporiazhenii Pravitel'stva*, 1915, no. 227, art. 1751.

105 *Otchet o deiatel'nosti Osobogo soveshchaniia i ob"edineniia meropriiatii po perevozke za sent. 1915-sent. 1916 g.* (Petrograd, 1916) p. 119.

106 'K nabliudatel'noi komissii o prichinakh zapozdaniia v ispolnenii nekotorykh zakazov priniat TsVPK-om' (24 May 1916), TsGVIA, f. 369, op. 4, d. 117, l. 504.

107 *Trudy vtorogo s"ezda*, pp. 109, 642; 'Doklad M.M. Poroshina o raskhodakh TsVPKa' (13 July 1916), TsGVIA, f. 369, op. 1, d. 206, l. 396.

108 'Otchet o deiatel'nosti TsVPKa', ibid., op. 4, d. 20, l. 28.

109 'Zhurnal zasedaniia biuro TsVPKa, no. 400' (3 Jan. 1917), TsGVIA, f. 13251, op. 1, d. 47, l. 6; Gorbachev, *Khoziaistvo i finansy VPKov*, p. 26.

110 'Pravila o zagotovleniiakh pri sredstve VPK-ov i glavnogo po snabzheniiu armii komiteta Vserossiiskikh zemskogo i gorodskogo soiuzov', *Izv. MVPKa*, no. 9-10, Nov. 1915, pp. 39-40.

111 *Izv. TsVPKa*, no. 18, 20 Oct. 1915, p. 2; ibid., no. 40, 11 Dec. 1915, p. 3.

112 'Summy zakazov i vydannye avansy' (20 Nov. 1915), TsGVIA, f. 369, op. 4, d. 20, ll. 78 obv.-79.

113 *Doklad schetnogo otdela na 26 sent. 1916 g.*, p. 21; 'Spravka po delu ob isproshennykh TsVPK-om avansakh', TsGVIA, f. 369, op. 4, d. 117, l. 706.

114 *Izv. TsVPKa*, no. 146, 17 Sept. 1916, p. 2; ibid., no. 159, 18 Oct. 1916, p. 3.

115 Korelin, 'Monopolii v metal. prom.', p. 184; F. Men'kov, 'Kontrol' za bankami', *Utro Rossii*, no. 126, 6 May 1916, p. 1, claims 'sixteen, eighteen, and even more' as the charge.

116 'Osobyi zhurnal Soveta ministrov' (23 July 1914), TsGIA, f. 1276, op. 10, d. 1201, l. 209.

117 *Ekonomicheskoe polozhenie Rossii nakanune Velikoi Oktiabr'skoi sotsialisticheskoi revoliutsii*, 3 vols (Moscow and Leningrad, 1957-67) vol. I, p. 108. A.I. Putilov, the bank's president, considered such operations 'quite advantageous and with limited risk' for the bank.

118 *Izv. TsVPKa*, no. 84, 19 Apr. 1916, p. 3.
119 *Sobr. uzak. i raspor.*, 1915, no. 369, art. 2775.
120 'Zapiska o dache TsVPK-om otzyvov o kredito-sposobnosti predpriiatii dlia polucheniia avansov bez obespecheniia', TsGAOR, f. 555, op. 1, d. 309, ll. 1-3.
121 'Lukomskii N.I. Bogatko' (24 March 1916), TsGVIA, f. 369, op. 6, d. 21, ll. 95-7; 'O komissionykh voznagrazhdeniiakh', ibid., op. 1, d. 279, ll. 1-4, 15, 29 obv., 37, 39-44.
122 This process is discussed in Tarnovskii, *Formirovanie*, pp. 229-53; Korelin, 'Monopolii v metal. prom.', pp. 223-65; and V.I. Bovykin and K.N. Tarnovskii, 'Kontsentratsiia pro-izvodstva i razvitie monopolii v metallo-obrabatyvaiushchei promyshlennosti Rossii', *Voprosy istorii*, no. 2 (1957) pp. 26-7.
123 Manikovskii, *Boevoe snabzhenie*, vol. II, p. 249.
124 'Doklad o deiatel'nosti Omskogo oblastnogo VPKa' (15 Aug. 1915), TsGAOR, f. 555, op. 1, d. 290, l. 6; 'Zapiska po obshchemu polo-zheniiu del vo Vladivostokskom VPKe', TsGVIA, f. 369, op. 1, d. 72, l. 69.
125 *Trudy vtorogo s"ezda*, p. 608.

NOTES TO CHAPTER 6: THE WAR-INDUSTRIES COMMITTEES AND THE REGULATION OF THE WAR ECONOMY

1 This only applied to railways in the rear. At the beginning of the war, one-third of the track and 35 per cent of the locomotives were placed under military authority. Multiple authority (*mnogovlastie*) in the rear survived until January 1917. See Sidorov, *Ekonomi-cheskoe polozhenie*, pp. 581-2, 598-628.
2 See ibid., pp. 46-9.
3 'Deviaty i ocherednoi S"ezd predstavitelei promyshlennosti i torgovli', *Doklad Soveta S"ezdov o merakh k razvitiiu proizvoditel'-nykh sil Rossii* (Petrograd, 1915) pp. 3-4; 'Rost'"gosudarstvennogo sotsializma" i otmena nalogov', *PT*, no. 4 (206), 23 Jan. 1916, pp. 85-90.
4 *Trudy ekonomicheskogo soveshchaniia, 3-4 ianvaria 1916 g.* (Moscow, 1916) p. 112.

5   *Narodnoe khoziaistvo v 1915 godu* (Petrograd,
     1918) p. 297. For a general study of the
     syndicate, see A.L. Tsukernik, 'Krovlia', *IZ*,
     vol. LII (1955) pp. 112–41.
6   *Monopolii v met. prom.*, pp. 157–8.
7   Darcy was a French citizen sent to Russia by
     the Banque de l'Union Parisienne in 1901. His
     father, Jean, was president of the French
     Comité des Forges. See Girault, *Emprunts
     russes et Investissements français*, p. 44.
8   *PT*, no. 23 (190), 26 Sept. 1915, pp. 361–2.
9   This committee, really an adjunct of the
     Association of Urals Mineowners, should not
     be confused with the Urals district com-
     mittee. The former had offices in Petrograd
     and represented large-scale industry, while
     the latter, with headquarters in Ekaterin-
     burg, dealt with medium and small-scale
     firms. See V.V. Adamov, 'Iz istorii mestnykh
     voenno-ekonomicheskikh organizatsii v gody
     pervoi mirovoi voiny (voenno-promyshlennye
     komitety na Urala)' in *Voprosy istorii Urala*
     (Sverdlovsk, 1958) pp. 82–95. Adamov, extra-
     polating from this unique bifurcation of
     responsibility, claims that the main function
     of the VPKs was to prevent the distribution
     of state orders to smaller firms.
10   'Metallurgicheskii otdel', 'Biuro po raspre-
     deleniiu metallov', TsGVIA, f. 369, op. 4, d.
     20, ll. 43 obv., 45–6.
11   The Metals Bureau continued to function after
     the establishment of the state's Metal-
     lurgical Committee, but was a 'powerless and
     superfluous organization' according to
     Tarnovskii, *Formirovanie*, p. 61. The same
     characterization is given in *Izv. Ob-va zav.
     i fab. Mosk. prom. raiona*, no. 1 (Jan. 1916)
     pp. 44–5.
12   'Zhurnaly soveshchaniia po ob"edineniiu
     deiatel'nosti predstavitelei soiuzov', *Izv.
     glav. kom. po snab. armii*, no. 17, 1 May
     1916, p. 158.
13   *Trudy vtorogo s"ezda*, pp. 37–8, 43.
14   A.P. Pogrebinskii, *Gosudarstvenno-monopoli-
     sticheskii kapitalizm v Rossii* (Moscow, 1959)
     p. 133; 'Predislovie', to D.I. Shpolanskii,
     *Monopolii ugol'no-metallurgicheskoi promy-
     shlennosti Iuga Rossii v nachale xx-veka*

(Moscow, 1953) p. 19.

15 *Monopolii v met. prom.*, pp. 189–90. From Myshlaevskii's farewell speech to the Metallurgical Committee delivered on 11 May 1917.

16 'Predlozhenie', 'Proekt polozheniia o metallicheskom komitete', 'Ob"iasnitel'naia zapiska k proektu', TsGVIA, f. 369, op. 1, d. 30, ll. 103–9.

17 *Otchet o deiatel'nosti otdelov TsVPKa na 1 noiabria 1915 g.* (Petrograd, 1916) pp. 34, 36.

18 TsGVIA, f. 369, op. 4, d. 20, l. 50. These may have constituted the obstacles to which Maidel referred.

19 'Ministr torgovli i promyshlennosti Goremykinu' (9/11 Nov. 1915), TsGVIA, f. 369, op. 1, d. 30, ll. 152–4.

20 For an account of the debate see *Russkie Vedomosti*, no. 271, 26 Nov. 1915, p. 2. For the statutes of the Metallurgical Committee see *Izv. MVPKa*, no. 13, Jan. 1916, pp. 33–4. Sidorov, *Ekonomicheskoe polozhenie*, pp. 148–9, regards the two proposals as a test of strength between Polivanov and Shakhovskoi.

21 *Monopolii v met. prom.*, pp. 191–2.

22 Pogrebinskii, 'Sindikat "Prodamet" v gody pervoi mirovoi voiny', pp. 25–6.

23 K.N. Tarnovskii, 'Komitet po delam metallurgicheskoi promyshlennosti i monopolisticheskie organizatsii', *IZ*, vol. LVII (1956) pp. 141–2.

24 *Novoe Vremia*, no. 14289, 19 Dec. 1915, pp. 4–5.

25 'Guzhon vo imeni Obshchestva zavodchikov i fabrikantov Moskovskogo raiona' (11 Nov. 1915), TsGVIA, f. 369, op. 16, d. 30, l. 77.

26 'Soveshchanie predstavitelei Zemgora', *Izv. MVPKa*, no. 21–2, May 1916, pp. 167–8; 'K predsedatel'iu Osobogo soveshchaniia' (May 1916), TsGVIA, f. 369, op. 1, d. 244, l. 30.

27 Zagorsky, *State Control of Industry*, p. 112; Tarnovskii, *Formirovanie*, pp. 85, 88–90.

28 For the TsVPK's position, see 'Konovalov Osobomu soveshchaniiu po oborone: dokladnaia zapiska o metallurgicheskom polozhenii' (13 Apr. 1916), TsGVIA, f. 369, op. 4, d. 117, l. 279. For the Unions' position, 'Osoboe soveshchanie po oborone upolnomochenemu po

delam metallurgii' (23 Feb. 1916), TsGVIA, f. 369, op. 1, d. 244, ll. 9-11; *Monopolii v. met. prom.*, p. 179.

29   *Izv. MVPKa*, no. 23-4, June 1916, pp. 156-71.

30   *Den'*, no. 143, 26 May 1916, pp. 2-3; *Rech'*, no. 143, 26 May 1916, p. 4. There were 23 resolutions in all amounting to a complete revision in regulatory practices. The resolutions are in *Zhurnaly TsVPKa*, no. 22, 13 June 1916; and 'Soveshchanie predstavitelei oblastnykh VPK-ov', *PT*, no. 22-3 (224), 11 June 1916, pp. 647-8.

31   *Izv. MVPKa*, no. 21-2, May 1916, p. 217.

32   'Zasedanie metallurgicheskogo otdela', *Izv. TsVPKa*, no. 108, 14 June 1916, pp. 2-3.

33   'Khronika', *PT*, no. 21 (223), 28 May 1916, p. 609; ibid., no. 41 (234), 15 Oct. 1916, p. 297.

34   'Zhurnal zasedaniia, no. 21' (10 May 1916), TsGVIA, f. 13251, op. 40, d. 30, l. 141.

35   *Den'*, no. 279, 10 Oct. 1916, p. 4; 'Regulirovanie metallicheskogo rynka', *Izv. TsVPKa*, no. 157, 13 Oct. 1916, p. 2; 'Khronika', *PT*, no. 42 (235), 23 Oct. 1916, p. 317.

36   'O kazennom predprinimatel'stve', TsGIA, f. 32, op. 1, d. 398, ll. 1-2.

37   Tarnovskii, *Formirovanie*, Table 10, p. 157.

38   On the decline in production, see *Monopolii v met. prom.*, pp. 576-7. The report to the Budget Commission, dated 6 August 1915, is quoted in Tarnovskii, *Formirovanie*, p. 166 citing TsGIA, f. 1278, op. 5, d. 447, ll. 270 obv.-271.

39   *Izv. TsVPKa*, no. 3, 2 Sept. 1915, p. 1. Ivanov, the chairman of the Foreign Purchase Committee, was also a member of the State Council as well as the board of the Kyshtym Mining Company, Russia's largest producer of electrolytic copper.

40   Quoted in Tarnovskii, *Formirovanie*, p. 167. The special commission was mooted in early August but only established on 28 September 1915 (see *Den'*, no. 220, 12 Aug. 1915, p. 6).

41   Bark's report to the Council of Ministers on the Anglo-Russian agreement is in 'Finansovye soveshchaniia soiuznikov vo vremia voiny', *KA*, no. 5 (1924) pp. 63-9. The ramifications of the agreement are discussed in Tarnovskii,

Formirovanie, pp. 166-80, and A.L. Sidorov, *Finansovoe polozhenie Rossii v gody pervoi mirovoi voiny* (Moscow, 1960) pp. 269-76.

42   'Osobyi zhurnal Soveta ministrov' (3 Nov. 1915), TsGVIA, f. 369, op. 1, d. 205, 1. 52. The resolution also called on the Ministry of Trade and Industry to investigate the high prices of orders.

43   *Izv. MVPKa*, no. 9-10, Nov. 1915, pp. 40-1.

44   'Ministr   torgovli   i   promyshlennosti Polivanovu' (12 Dec. 1915), TsGVIA, f. 369, op. 16, d. 1, 11. 1-2. According to the committee's statutes, the public organizations enjoyed a nine to five edge over ministerial representatives.

45   'Ministr finansov Polivanovu' (15 Dec. 1915), ibid., 11. 4-5.

46   'Metalsnabzhenie Polivanovu' (3 Mar. 1916), ibid., op. 1, d. 120, 11. 124-8.

47   'Metalsnabzhenie Polivanovu', ibid., d. 244, 11. 12-18.

48   'Pravlenie O-va Mednoprokatnogo i truboprokatnogo zavoda byvsh. Rozenkrantsa: zapiska po voprosu o snabzhenii zavoda metallami', ibid., d. 120, 11. 166-9; *Monopolii v met. prom.*, pp. 518-23; and Tarnovskii, *Formirovanie*, p. 183, who claims that the accusations against Ivanov were false.

49   *Outline of Activities of the Central War Industrial Committee*, p. 4; 'Zhurnal zasedaniia biuro TsVPKa, no. 17' (27 Aug. 1915), TsGVIA, f. 13251, op. 1, d. 39, 11. 103, 106. Bakhmet'ev later became the Provisional Government's ambassador in Washington. Morozov was the Moscow VPK's representative. The Russian Government Supply Committee was set up in October 1914 by the Russian embassy to organize purchases. However, according to a report filed by an agent of the Ministry of Trade and Industry, as of July 1915 it performed only 'registrational-archival' functions. Its chairman was spending the summer months in Nantucket Island and the committee had not met since May. 'Vypiska iz doneseniia agenta   Ministerstva   torgovli   i   promyshlennosti v Amerike' (22 July/4 Aug. 1915), TsGVIA, f. 369, op. 1, d. 89, 11. 10-14.

50   National   Archives   Division   (Washington,

D.C.), Record Group 261 (Russian Government
Supply Committee in the United States), L6,
'Documents related to minutes of Supply Committee', pp. 30-3; L37, part I, vol. 1, pp.
14-31; and L37, part II, vol. 3, 'Otchet o
deiatel'nosti otdela TsVPKa', pp. 3-6. The
members of the Supply Committee were extremely   ambivalent   about   Morgan's   role.
Sapozhnikov resented it; Bakhmet'ev claimed
there were both advantages and disadvantages;
and Morozov wrote to the TsVPK that 'I do not
like him, nor do I rate him very highly'. For
Morozov's letter see 'Glavnoupol'nomochennyi
TsVPKa v Amerike Guchkovu' (5 Jan. 1916),
TsGVIA, f. 369, op. 4, d. 117, ll. 117-20.

51   'O rabote nashikh organizatsii v Londone i
Parizhe', TsGVIA, f. 369, op. 1, d. 20, l.
28. For a thorough study of the Anglo-Russian
Committee see D.S. Babichev, 'Deiatel'nost'
russkogo   pravitel'stvennogo   komiteta   v
Londone', pp. 276-92.

52   'Polivanov   Shtiurmeru'   (15   Feb.   1916),
TsGVIA, f. 369, op. 1, d. 205, l. 117; ibid.
(25 Feb. 1916), ll. 154-5.

53   V.N. Shakhovskoi, *Sic Transit Gloria Mundi*
(Paris, 1952) p. 106.

54   'Osobyi zhurnal Soveta ministrov' (13 Mar.
1916), TsGVIA, f. 369, op. 1, d. 184, l. 224
obv.

55   This   is   Polivanov's   explanation   of   his
dismissal,   given   in   *Padenie   tsarskogo
rezhima*, vol. III, p. 205. G. Katkov, *Russia
1917: The February Revolution*, 2nd edn
(London, 1969) pp. 212-17, 258, stresses
Polivanov's ties with Guchkov and the consternation   they   caused   the   royal   couple.
Semennikov, *Monarkhiia pered krusheniem*, p.
114, considers Polivanov's handling of the
Putilov Works' sequestration an important
factor.

56   'Zhurnal zasedaniia biuro TsVPKa, no. 288'
(28 July 1916), TsGVIA, f. 13251, op. 1, d.
45, l. 54 obv.; *Izv. TsVPKa*, no. 129, 6 Aug.
1916, p. 1.

57   *Aktsionerno-paevye   obshchestva*,   p.   105.
Rabinovich later served the Soviet state as
head of the coal industry (1920-7), only to
be arrested with other engineers in the

Shakhty affair. He received a six-year prison sentence. See Bailes, *Technology and Society under Lenin and Stalin*, pp. 90-4.

58 *Monopol. kapital v neft. prom.*,Tables 1 and 2 (pp. 511, 514a), pp. 125, 214, 490, 515.

59 Volobuev, 'Politika proizvodstva ugol'nykh i neftianykh monopolii', p. 72, states that 'on the eve of the First World War only nineteen firms remained in Produgol' producing one-half of the coal in the Donets'. See also 'Produgol'', *KA*, no. 18 (1926) pp. 140-5, for evidence of its decline.

60 *Izvestiia Osobogo soveshchaniia po toplivu*, no. 2 (1917) pp. 136, 138. The deficit resulting from cessation of imports has been variously estimated at 322 and 450 million puds. For the latter figure see L.B. Kafengauz, 'Snabzhenie strany mineral'nym toplivom vo vremia voiny', *Trudy komissii po izucheniiu sovremennoi dorogovizny*, 4 vols (Moscow, 1915) vol. II, p. 138.

61 Sidorov, *Ekonomicheskoe polozhenie*, Table 32 (p. 549); 'Spravka ot Ministra torgovli i promyshlennosti' (18 Nov. 1915), TsGIA, f. 1276, op. 11, d. 888, ll. 178, 183-4.

62 *Gornozavodskoe delo*, no. 1, 8 Jan. 1915, p. 10217; *Den'*, no. 60, 3 Mar. 1915, p. 6; ibid., no. 67, 20 Mar. 1915, p. 2; no. 71, 14 Mar. 1915, p. 3.

63 *Den'*, no. 87, 31 Mar. 1915, p. 6; *Russkoe Slovo*, no. 75, 3 Apr. 1915, p. 3.

64 *Russkoe Slovo*, no. 221, 27 Sept. 1915, p. 5; ibid., no. 224, 1 Oct. 1915, p. 4; no. 236, 15 Oct. 1915, p. 4. The producers' plan was presented by I.A. Knotte, Avdakov's successor as chairman of Produgol'.

65 *Izv. TsVPKa*, no. 11, 1 Oct. 1915, p. 2.

66 *Izv. MVPKa*, no. 4-5, Oct. 1915, p. 17; *Russkoe Slovo*, no. 224, 1 Oct. 1915, p. 4.

67 'Rekvizitsiia topliva', *Izv. MVPKa*, no. 9-10, Nov. 1915, p. 75; *Russkoe Slovo*, no. 271, 26 Nov. 1915, p. 5.

68 Among those voting in the majority for a general requisition were the representatives of the Zemstvo Union and the Union of Towns. The latter's economic council later criti-cized the TsVPK for siding with the mine-owners against the interests of the general

population. See *Deiatel'nost' ekonomicheskogo otdela Vserossiiskogo Soiuza Gorodov* (Petrograd, 1916) p. 12. *Russkie Vedomosti*, no. 297, 29 Dec. 1915, p. 2, also favoured the state monopoly.

69   S. Kornfeld', 'Gosudarstvennaia monopoliia torgovli kamennougol'nym toplivom', *Izv. TsVPKa*, no. 140, 1 Sept. 1916, p. 3.

70   'K istorii proekta "Tsentrougolia"', *Izv. Osopot*, no. 3 (Apr. 1917) pp. 48-53 contains the protocols of a meeting of industrialists held on 31 October 1916; 'Tsentrougol'', *Izv. TsVPKa*, no. 171, 19 Nov. 1916, pp. 3-4; 'Anketa o Tsentrougle', ibid., no. 172, 24 Nov. 1916, pp. 1-2.

71   *Den'*, no. 346, 16 Dec. 1916, p. 4.

72   Zagorsky, *State Control of Industry*, pp. 214-22, Appendix V (pp. 285-92). This form of state control, which was also applied to other commodities, constituted the basis for the Soviet state's 'centres' and *glavki* during 'War Communism' and the early NEP period. 'TsVPK predsedateliu Osobogo soveshchaniia' (11 Apr. 1916), TsGVIA, f. 369, op. 21, d. 40, l. 15; see also *Monopol. kapital v neft. prom.*, pp. 128-9, which contains the minutes of the TsVPK's fuel section for 10 May 1916. At this meeting, it was resolved to support the petition for state subsidies filed by 'Grumant', a coal company with mines in Spitzbergen, and to sponsor an expedition for investigating the possibility of wider exploitation.

73   As far as alternative sources were concerned, the TsVPK encouraged peat production and allocated 15,000 rubles for experiments with shale deposits in the Baltic area. See *Izv. glav. kom. po snab. armii*, no. 22, 15 July 1916, p. 126.

74   *Vidy na snabzhenie strany toplivom v 1916 godu* (Petrograd, 1916) p. 3.

75   *PT*, no. 25 (192), 10 Oct. 1915, pp. 433-4.

76   'Tseny na neft'', *Russkoe Slovo*, no. 76, 4 Apr. 1915, p. 4.

77   *Den'*, no. 21, 22 Jan. 1916, p. 6; *Monopol. kapital v neft. prom.*, p. 476.

78   *Monopol. kapital v neft. prom.*, pp. 88-9, 476.

79 Sidorov, *Ekonomicheskoe polozhenie*, p. 529.

80 *Izv. TsVPKa*, no. 57, 8 Feb. 1916, p. 1.

81 'Predel'naia tsena na neft'', *Izv. glav. kom. po snab. armii*, no. 15, 15 Mar. 1916, p. 161.

82 *Izv. MVPKa*, no. 19-20, Apr. 1916, p. 30.

83 *Monopol. kapital v neft. prom.*, pp. 212-22; 'Polozhenie Bakinskoi neftianoi promy-shlennosti', *Izv. TsVPKa*, no. 182, 17 Dec. 1916, pp. 2-4, contains the debate which followed Gukasov's report. One is struck by the similarity between these arguments and those advanced by the Zemstvo Union with respect to food prices. See Thomas Fallows, 'Politics and the War Effort in Russia: the Union of Zemstvos and the Organization of the Food Supply, 1914-1916', *Slavic Review*, vol. XXXVII (1978) pp. 88-9.

84 *PT*, no. 7 (251), 18 Feb. 1917, p. 172; Sidorov, *Ekonomicheskoe polozhenie*, p. 527, citing *Izv. Osopot*, no. 5 (1917) p. 159.

85 *Izv. TsVPKa*, no. 182, 17 Dec. 1916, pp. 2-4.

86 *Den'*, no. 4, 5 Jan. 1917, p. 5.

87 *Monopol. kapital v neft. prom.*, pp. 227-80, contains the minutes of the Council of Minis-ters' session. For the new prices see *Monopol. kapital v neft. prom.*, p. 496.

88 *Trudy vtorogo s"ezda*, p. 418. The delegate was A.I. Churkin. Samara, like other cities on the Volga, was dependent on oil for grain shipping.

89 Gos. Duma, IV Sozyv: *Sten. otchety*, ses. 4, zas. 24 (19 Feb. 1916) cols 1936-42.

90 *Monopol. kapital v neft. prom.*, pp. 139-41.

91 *Izv. TsVPKa*, no. 106, 9 June 1916, p. 2.

92 *Izv. glav. kom. po snab. armii*, no. 22, 15 July 1916, p. 188.

93 Gos. Duma, IV Sozyv: *Sten. otchety*, ses. 1, zas. 33 (29 Mar. 1913) cols 233-6.

94 S. Kornfeld', 'O sozdanii gosudarstvennoi neftianoi promyshlennosti na Apsheronskom poluostrove', *Izv. TsVPKa*, no. 124, 26 July 1916, p. 1.

95 *Monopol. kapital v neft. prom.*, pp. 483-4.

96 *Narodnoe khoziaistvo v 1915 godu*, p. 142.

97 Zagorsky, *State Control of Industry*, Appendix XXIV (p. 340); *Narodnoe khoziaistvo v 1915 godu*, p. 143. Prices are for first class cotton as delivered to Moscow by rail.

98   'Manufakturnyi rynok', *Izv. Ob-va zav. i fab.
      Mosk. prom. raiona*, no. 8 (1915) p. 11. P.P.
      Riabushinskii, chairman of the Society of
      Cotton Manufacturers, and N.D. Morozov voted
      against its decision to ask for state inter-
      vention. The view that the state had no right
      to interfere was attributed to the former by
      Kommercheskii telegraf, no. 661, 8 June 1915,
      p. 3.
99   The statutes of the committee, hereafter the
      Committee on Cotton, are in *Osobye sove-
      shchaniia i komitety voennogo vremeni* (Petro-
      grad, 1917) pp. 79-83.
100   'Protokol pervogo zasedaniia komiteta' (13
      Aug. 1915), TsGIA, f. 1131, op. 1, d. 6, 1.
      1.
101   'Khlopkovyi rynok', *Izv. Ob-va zav. i fab.
      Mosk. prom. raiona*, no. 5-6 (1916) p. 19;
      'Protokol 11-ogo zasedaniia komiteta' (17
      Oct. 1915), TsGIA, f. 1131, op. 1, d. 6, 1.
      83.
102   Quoted in FO 371 2454 N117006: 'Acting
      Consul-General Lockhart to Buchanan', 8/21
      July 1915.
103   N. Pokrovskii, 'Voennye pribyli glavneishikh
      otraslei   tekstil'noi   promyshlennosti',
      Vestnik finansov, promyshlennosti i torgovli,
      no. 21 (1917) pp. 292-3.
104   Ibid., p. 294.
105   Participation of the VPKs in the regulation
      of cotton extended to the Erevan VPK which
      was commissioned 'to discuss the business of
      requisitions and execute it with the prior
      approval of the Committee on Cotton' and 'to
      supervise the purchase of cotton at the fixed
      price' ('Protokol 15-ogo zasedaniia komiteta'
      (11 Dec. 1915), TsGIA, f. 1131, op. 1, d. 6,
      1. 120).
106   *Sobranie   uzakonenii   i   rasporiazhenii
      Pravitel'stva*, 1916, no. 2, art. 4. An
      English translation of the decree is in
      Zagorsky, *State Control of Industry*, Appendix
      VI (pp. 293-6).
107   'Protokol 17-ogo zasedaniia komiteta' (13
      Jan. 1916), TsGIA, f. 1131, op. 1, d. 6, 11.
      147-9 (emphasis mine).
108   V.Ia. Laverychev, *Monopolisticheskii kapital
      v tekstil'noi promyshlennosti Rossii, 1900-*

1917 (Moscow, 1963) p. 264. Laverychev asserts that the delay was crucial because 'massive sales of yarn' were conducted in January and February.

109   *Izv. MVPKa*, no. 19-20, Apr. 1916, p. 196.

110   Quoted in Zagorsky, *State Control of Industry*, p. 134.

111   *Izv. Ob-va zav. i fab. Mosk. prom. raiona*, no. 12 (1916) p. 21.

112   For a description of conflict between spinners and weavers, see Laverychev, *Monopol. kapital v tekst. prom.*, pp. 268-9. See P. Tobolkin, 'Normirovka khlopchatobumazhnykh tkanei', *Izv. Ob-va zav. i fab. Mosk. prom. raiona*, no. 7 (1916) p. 20, for a justification of price increases of up to 125 per cent for fabrics.

113   'Protokol 20-ogo zasedaniia komiteta' (3 Mar. 1916), TsGIA, f. 1131, op. 1, d. 6, ll. 169-73. Cotton was distributed to 73 enterprises.

114   'Protokol 23-ego zasedaniia komiteta' (26 May 1916), ibid. l. 189; 'Deiatel'nost' khlopchatobumazhnogo otdela', *Izv. MVPKa*, no. 27-30, Aug.-Sept. 1916, pp. 78-9. What an ironic situation!

115   *Izv. MVPKa*, no. 27-30, Aug.-Sept. 1916, p. 79.

116   Laverychev, *Monopol. kapital v tekst. prom.*, pp. 318-20 was the first to make this case. The economic section of the Union of Towns accepted the cotton section's figures of 66+ per cent for April to November 1916. See *Polozhenie tekstil'noi promyshlennosti* (Moscow, 1917) p. 9. Sidorov, *Ekonomicheskoe polozhenie*, p. 385 accepts this figure as well.

117   'Protokol 20-ogo zasedaniia komiteta' (3 Mar. 1916), TsGIA, f. 1131, op. 1, d. 6, l. 169.

118   Laverychev, *Monopol. kapital v tekst. prom.*, p. 320; *Deiatel'nost' oblastnykh i mestnykh VPK-ov*, vol. I, p. 41.

119   'Zhurnal zasedaniia biuro' (30 Sept. 1916), Tsentral'nyi Gosudarstvennyi Arkhiv goroda Moskvy (hereafter TsGAGM), f. 1082, op. 1, d. 119, ll. 6-9.

120   'Zhurnal biuro Moskovskogo VPKa' (7 Oct. 1916), ibid., l. 38.

121   Laverychev, *Monopol. kapital v tekst. prom.*,

p. 318 estimates that in 1916 50 per cent of textile production went towards the army but this figure includes Petrograd's mills.

122 This figure is given by Laverychev, *Monopol. kapital v tekst. prom.*, p. 319.

123 Based on the cotton section's own investigation quoted in Laverychev, *Monopol. kapital v tekst. prom.*, p. 268. The undelivered yarn amounted to 358,000 puds and was sold on the market at an average price of 78 rubles per pud compared to the army's price of 43 rubles per pud.

124 'Protokol 48-ogo zasedaniia komiteta dlia zavedyvaniia snabzhenii syrev khlopchato-bumazhnykh fabrik' (27 Jan. 1917), TsGAGM, f. 1082, op. 1, d. 131, l. 2 obv.

125 'Sovet Obshchestva fabrikantov khlopchato-bumazhnoi promyshlennosti k Ministru torgovli i promyshlennosti: dokladnaia zapiska' (Jan. 1917), TsGIA, f. 1131, op. 1, d. 26, ll. 214-17.

126 *Otchet o deiatel'nosti Osobogo soveshchaniia po perevozke,* pp. 174-5, 179.

127 For an account critical of the administration of evacuation see A.L. Sidorov, 'Evakuatsiia russkoi promyshlennosti vo vremia pervoi mirovoi voiny', *Voprosy istorii*, no. 6 (1947) pp. 3-25.

128 'Evakuatsiia predpriiatii', *Izv. MVPKa*, no. 8, Nov. 1915, p. 42.

129 For an account of how the refugee problem was handled, see Gleason, 'The All-Russian Union of Towns and the All-Russian Union of Zemstvos', pp. 138-78.

130 Of 233 industrial enterprises evacuated from Vilna, Minsk and Dvinsk, 67 went to Moscow, 50 to central Russia (Volga basin), 47 to South Russia, 7 to Siberia, and 62 elsewhere. *Izv. MVPKa*, no. 17-18, Mar. 1916, p. 62. According to Sidorov, *Ekonomicheskoe polozhenie*, p. 223, of 395 enterprises evacuated from Riga, Revel and Windau, 121 wound up in the Moscow area.

131 A. Urin, 'Evakuatsiia i kontrol' nad fabrikami i zavodami', *Izv. MVPKa*, no. 3, Sept. 1915, p. 2.

132 *Otchet o deiatel'nosti Osobogo soveshchaniia po perevozke,* p. 46; *Izv. MVPKa*, no. 6-7,

Oct. 1915, p. 83.

133   A. Urin, 'Deiatel'nost' komissii po obsledov-
      aniiu   promyshlennykh   predpriiatii   i   po
      evakuatsionnym voprosam', *Izv. MVPKa*, no. 12,
      Dec. 1915, p. 7.

134   The Committee was careful about which pet-
      itions from Jews it supported. Thus, for
      example, the petition of the owners of 'Mazur
      and Shnurkin' was supported but that of Sh.M.
      Lampert, a 'contractor', was rejected. The
      second congress was more liberal, calling for
      the abolition of all restrictions to the
      settlement of nationalities. *Izv. MVPKa*, no.
      8, Nov. 1915, p. 58; ibid., no. 9-10, Nov.
      1915, pp. 72-3.

135   *Trudy s"ezda predstavitelei VPK-ov*, p. 175;
      Izv. MVPKa, no. 6-7, Oct. 1915, p. 60.

136   *Sobranie   uzakonenii   i   rasporiazhenii
      Pravitel'stva*, 1915, no. 288, art. 2135.

137   *Izv. MVPKa*, no. 9-10, Nov. 1915, p. 60.

138   *Izv. TsVPKa*, no. 22, 29 Oct. 1915, p. 4;
      Sidorov, *Ekonomicheskoe polozhenie*, p. 604;
      and Pogrebinskii, *Gosudarstvenno-monopolis-
      ticheskii kapitalizm v Rossii*, pp. 145-6.

139   Sidorov, *Ekonomicheskoe polozhenie*, pp. 602,
      607.

140   *Izv. MVPKa*, no. 25-6, July 1916, p. 155;
      ibid., no. 8, Nov. 1915, p. 46. Riabushinskii
      also called for the public supervision of
      food supplies and the creation of war-
      commercial committees, neither of which came
      to anything.

141   *Russkie Vedomosti*, no. 268, 21 Nov. 1915, p.
      6. Trepov broke a deadlocked vote in the
      Special Council for Transport by deciding
      against representation for the public organ-
      izations. For the work of the Provisional
      Regulatory Committee see A.L. Sidorov,
      'Zheleznodorozhnyi transport Rossii v pervoi
      mirovoi voine i obostrenie ekonomicheskogo
      krizisa v strane', *IZ*, vol. XXVI (1948) pp.
      49-51.

142   'Obshchestvennyi kontrol' nad zheleznymi
      dorogami', *Izv. MVPKa*, no. 15-16, Feb. 1916,
      pp. 20-4. The Soviet practices of inspections
      (*smotry*) and 'light cavalry raids' bears
      striking resemblance to this plan.

143   'Obshchestvennyi kontrol'', *Izv. TsVPKa*, no.

79, 2 Apr. 1916, p. 4.

144 *Izv. TsVPKa*, no. 101, 28 May 1916, p. 2.

145 Zagorsky, *State Control of Industry*, p. 51; Tsent. stat. upr., *Rossiia v mirovoi voine*, Tables 2-6 (pp. 18-22).

146 'Sovet s"ezdov predstavitelei promyshlennosti i torgovli Petrogradskomu Obshchestvu fabrikantov i zavodchikov' (4 June 1915), TsGIA, f. 150, op. 1, d. 207, 1. 194. The Association of Industry and Trade passed on information received from Headquarters.

147 'Dokladnaia zapiska Guzhona k otdelu promyshlennosti' (10 June 1915), ibid., 1. 203; *Trudy s"ezda predstavitelei VPK-ov*, p. 156.

148 *Izv. MVPKa*, no. 25-6, July 1916, p. 212.

149 Ibid., no. 23-4, June 1916, p. 212.

150 *Trudy vtorogo s"ezda*, p. 633.

151 'Zhurnal zasedaniia Soveta' (11 Aug. 1915), TsGIA, f. 150, op. 1, d. 60, 1. 260.

152 'Nekotorye soobrazheniia TsVPKa o poriadke predstavlennia otsrochek po prizyvu voenno-obiazannym rabochim i sluzhashchim promyshlennykh predpriiatii' (24 Sept. 1915), TsGAOR, f. 12642, op. 2, d. 34, 11. 113-19.

153 The statutes are in *Izv. MVPKa*, no. 14, Jan. 1916, pp. 32-8.

154 *Trudy s"ezda predstavitelei VPK-ov*, pp. 163, 166, 216, 268-71. According to a decree of 9 March 1915, women and children at least fifteen years of age could be employed for night-work and underground so long as the next work day did not begin before noon. See *Sobranie uzakonenii i rasporiazhenii Pravitel'stva*, 1915, no. 90, art. 759. According to *Izv. TsVPKa*, no. 25, 5 Nov. 1915, p. 2, 'Thanks to their physical stamina, comparatively cheap wages, moderate demands in terms of housing and food, sobriety, and lack of crime, the Chinese pay back with interest the employer who covers the cost for transporting and sustaining them.' For a general discussion of the question see L. Siegelbaum, '"Another Yellow Peril": Chinese Migrants in the Russian Far East and the Russian Reaction before 1917', *Modern Asian Studies*, vol. XII (1978) pp. 307-30, esp. 326-8.

155 *Sobranie uzakonenii i rasporiazhenii*

Pravitel'stva, 1915, no. 311, art. 2312; Kir'ianov, *Rabochie Iuga Rossii*, p. 42. For a report on the recruitment of Chinese workers in Kharbin, see *Doklad Soveta s"ezdov XLI gornopromyshlennikov Iuga Rossii* (Khar'kov, 1916) pp. 8-12.

156 Kir'ianov, *Rabochie Iuga Rossii*, p. 39; *Materialy k uchetu rabochego sostava i rabochego rynka*, 2 vols (Petrograd, 1916-17) vol. I, p. 119; vol. II, pp. 116, 120; *Izv. Osopot*, no. 3 (1917) p. 2.

157 'Povyshenie produktivnosti truda', *Utro Rossii*, no. 194, 16 July 1915, p. 1.

158 'Trekhsmennaia rabota mekhanicheskikh stankov', *Izv. MVPKa*, no. 2 (Sept. 1915) pp. 5-7. This was based on 'Doklad o pervoocherednykh zadachakh pravitel'stva i obshchestva po gosudarstvennoi oborne, A. Konshin' (4 Aug. 1915), TsGVIA, f. 13251, op. 11, d. 23, ll. 1-6.

159 *Trudy s"ezda predstavitelei VPK-ov*, p. 206. The congress resolved that enterprises and not individual workers should work 360 days a year (p. 212).

160 For the law see Feldman, *Army, Industry, and Labor in Germany*, Appendix, pp. 535-41; for the concept, R.B. Armeson, *Total Warfare and Compulsory Labor: A Study of the Military-Industrial Complex in Germany during World War I* (The Hague, 1964) chs 3 and 4.

161 Laverychev, *Tsarizm i rabochii vopros*, p. 284.

162 The plans for militarization are discussed in ibid., pp. 270-94, and B.B. Grave, 'Militarizatsiia promyshlennosti i rossiiskii proletariat v gody pervoi mirovoi voiny', in *Iz istorii rabochego klassa i revoliutsionnogo dvizheniia*, *Sbornik* (Moscow, 1958) pp. 405-26.

163 For conversation between Stuermer and Albert Thomas see Paleologue, *An Ambassador's Memoirs*, vol. II, p. 252. The views of Polivanov and Khvostov are in *Prologue to Revolution: Notes of A.N. Iakhontov on the Secret Meetings of the Council of Ministers, 1915*, ed. M. Cherniavsky (Englewood Cliffs, N.J., 1967) p. 146.

164 *Organizatsiia voenno-promyshlennykh komitetov*

(Petrograd, 1915) p. 4. This publication was prepared for the first congress by the TsVPK.

165   *Trudy vtorogo s"ezda*, p. 3.

NOTES TO CHAPTER 7:   THE LABOUR QUESTION

1   See, for example, M.T. Florinsky, *The End of the Russian Empire* (New York, 1961) pp. 163-5; M. Ferro, *La Revolution de 1917* (Paris, 1967) pp. 58-9; Katkov, *Russia 1917*, pp. 47-56, 147-8, 317-20; M. Liebman, *The Russian Revolution*, trans. A.J. Pomerans (London, 1970) pp. 87-8; and J.L.H. Keep, *The Russian Revolution: A Study in Mass Mobilization* (New York, 1976) pp. 53-7, 557.

2   The term is used explicitly by Laverychev, *Tsarizm i rabochii vopros*, p. 300, and I.P. Leiberov, 'Petrogradskii proletariat v gody pervoi mirovoi voiny', in *Istoriia rabochikh Leningrada*, vol. I, p. 490. See also L.S. Gaponenko, *Rabochii klass Rossii v 1917 godu* (Moscow, 1970) pp. 129-36, and Borisov, *Bor'ba bol'shevikov protiv voenno-promyshlennykh komitetov*, passim.

3   Kir'ianov, *Rabochie Iuga Rossii*, pp. 223-31.

4   *Trudy s"ezda predstavitelei VPK-ov*, pp. 194, 200, 204, 206, 209-11.

5   For Chelnokov's view, see *Burzhuaziia nakanune*, p. 34. See also B.B. Grave, *K istorii klassovoi bor'by v Rossii v gody imperialisticheskoi voiny* (Moscow and Leningrad, 1926) p. 139.

6   *Russkoe Slovo*, no. 160, 12 July 1915, p. 6. Prince L'vov addressed workers as 'our sons' and asked them to arrange their work patterns so that not one minute would be wasted.

7   V. Skal'tsov, 'K voprosu ob uchastii rabochikh v TsVPKe', *Voprosy Strakhovaniia*, no. 7 (45) (1915) p. 3.

8   S.V. Tiutiukin, *Voina, mir, revoliutsiia. Ideinaia bor'ba v rabochem dvizhenii Rossii, 1914-1917 gg.* (Moscow, 1972) p. 206.

9   *Prologue to Revolution*, p. 173.

10   V.N. Zalezhskii, *Iz vospominanii podpol'shchika* (Khar'kov, 1931) p. 131. Zalezhskii admits that 'the entire election campaign proceeded virtually freely and without mass

arrests'.
11 'OK R.S.D.R.P. listok', TsGAOR, f. 102 (1915), d. 347, l. 7.
12 *Izvestiia zagranichnogo sekretariata OK R.S.D.R.P.* no. 3, 5 Feb. 1916 quoted in B. Dvinov, *Pervaia mirovaia voina i Rossiiskaia Sotsial-demokratiia* (Inter-University Project on the History of the Menshevik Movement, no. 10, New York, 1962) p. 74.
13 'Deklaratsiia', quoted in ibid., pp. 74-5. See also A.A. Abramov, 'Bor'ba bol'shevist-skikh organizatsii protiv sotsial-shovinizma i tsentrizma v Rossii (1914-fev. 1917)', *Voprosy istorii KPSS*, no. 11 (1963) pp. 52, 55.
14 'Antivoennaia rabota bol'shevikov v gody pervoi mirovoi voiny, 1914-1917 gg.', *Istoricheskii Arkhiv*, no. 5 (1961) pp. 85-8. The Petersburg Committee's instructions are discussed in I.P. Leiberov, 'O vozniknovenii revoliutsionnoi situatsii v Rossii v gody pervoi mirovoi voiny (iiul'-sent. 1915 g.)', *Istoriia SSSR*, no. 6 (1964) pp. 50-1.
15 'Dokumenty biuro TsK R.S.D.R.P. v Rossii, iiul' 1914-fevral' 1917 gg.', *Voprosy istorii KPSS*, no. 8 (1965) p. 93. Lenin referred to the Petersburg Committee's plan as 'an illusion' in a letter to Shliapnikov. See *Polnoe sobranie sochinenii*, vol. XLIX, pp. 159-60.
16 'Doklad Ministru vnutrennykh del' (25 Sept. 1915), TsGAOR, f. 102 (1915), d. 347, ll. 14-19. Shliapnikov wrote to Lenin on 30 November that 'at several factories Mensheviks were elected but they were given Bolshevik instructions....There were also instances where Bolsheviks were given Menshevik instructions.' He concluded that 'the attitude of the workers is not easy to make out'. Quoted in K.F. Sidorov, 'Rabochee dvizhenie v Rossii v gody imperialisticheskoi voiny, 1914-1917 gg.', in *Ocherki po istorii Oktiabr'skoi revoliutsii*, ed. M.N. Pokrovskii, 2 vols (Moscow and Leningrad, 1927) vol. I, p. 409.
17 *Den'*, no. 274, 5 Oct. 1915, p. 3. Zalezhskii also impersonated a non-party elector.
18 The fullest accounts of the meeting are in A.G. Shliapnikov, *Kanun semnadtsatogo goda:*

Vospominaniia   i   dokumenty   o   rabochem dvizhenii i revoliutsionnom podpol'e za 1914-1916 gg. (Moscow, 1923) pp. 112-35; and 'Doklad Ministru vnutrennykh del' (30 Nov. 1915), TsGAOR, f. 102 (1915), d. 347, ll. 133-46.

19   Quoted in Shliapnikov, *Kanun semnadtsatogo goda*, pp. 135-6.

20   *Rabochee dvizhenie v Petrograde v 1912-1917 gg., Dokumenty i materialy* (Leningrad, 1958) pp. 362-3. The police report is dated 5 December 1915.

21   TsGAOR, f. 102 (1915), d. 347, l. 141.

22   Shliapnikov, *Kanun semnadtsatogo goda*, p. 118.

23   Katkov, *Russia 1917*, pp. 52-5.

24   Quoted in Shliapnikov, *Kanun semnadtsatogo goda*, pp. 128-30.

25   TsGAOR, f. 102 (1915), d. 347, ll. 133-4.

26   See *Russkie Vedomosti*, no. 245, 25 Oct. 1915, p. 5; Kir'ianov, *Rabochie Iuga Rossii*, pp. 123, 157, 166; and TsGAOR, f. 102 (1915), d. 347 prilozhenie, ll. 48, 51, 53, 56, 63, 72, 83, 106.

27   *Russkie Vedomosti*, no. 261, 13 Nov. 1915, p. 4, and *Izv. MVPKa*, no. 8, Nov. 1915, pp. 43-4. Returns from other cities are scarce and provide little basis for comparison. In Kiev, where a workers' group was elected, 62.8 per cent of those eligible cast votes. In Khar'kov, where the boycottists proved victorious, only 35 per cent voted, while it was reported that less than half of Nikolaev's workers (not all of whom were eligible) participated in the election which resulted in three representatives being sent to the War-Industries Committee. See *Izv. TsVPKa*, no. 159, 18 Oct. 1916, p. 3, and V.L. Kharitonov, 'Bor'ba bol'shevikov Ukrainy za boikot voenno-promyshlennykh komitetov', *Nauchnye doklady vyshei shkoly istoricheskoi nauki*, no. 4 (1958) pp. 63-6.

28   'Nizhegorodskii gubernskii nachal'nik direktoru D.P.', TsGAOR, f. 102 (1915), d. 347 prilozhenie, 1. 53; *Izv. MVPKa*, no. 17-18, Mar. 1916, p. 170; Ia. Bazanov, 'Sem' mesiatsev partiinoi raboty v Khar'kove', *Proletarskaia revoliutsiia*, no. 9 (1922) pp. 91-4,

122; S.D. Chkhartishvili, 'Bor'ba bol'shevi-kov zakavkaz'ia protiv imperialisticheskoi voiny', in *Pervaia mirovaia voina*, ed. A.L. Sidorov (Moscow, 1968) p. 318.

29   The votes were 69-21 in Moscow, 172-16 in Kiev, 45-25 in Samara, and 52-31 in Omsk. Sources are respectively *Izv. MVPKa*, no. 8, Nov. 1915, p. 43; Kharitonov, 'Bor'ba bol'shevikov Ukrainy', p. 63; N.F. Panov, 'Moia rabota v Samare', *Krasnaia byl'*, no. 1 (1922) pp. 5, 8; TsGAOR, f. 102 (1915) d. 347 prilozhenie, l. 207.

30   Kir'ianov, *Rabochie Iuga Rossii*, pp. 123-7; *Izv. MVPKa*, no. 19-20, Apr. 1916, p. 212; *Den'*, no. 70, 12 Mar. 1916, p. 6.

31   TsGAOR, f. 102 (1915), d. 347 prilozhenie, l. 95; *Izv. MVPKa*, no. 19-20, Apr. 1916, p. 212; *Izv. TsVPKa*, no. 151, 29 Sept. 1916, p. 5.

32   *Izv. TsVPKa*, no. 178, 8 Dec. 1916, p. 2.

33   T. Dan, *The Origins of Bolshevism*, trans. J. Carmichael (New York, 1970) p. 397; for Zin-oviev's characterization, see 'Vpolzli v legal'nost'', *Sotsial Demokrat*, no. 50 (1916), quoted in K. Kareev, 'Rabochaia pechat' i voenno-promyshlennye komitety', *Krasnaia letopis'*, no. 21 (1926) p. 150.

34   See 'K istorii Gvozdevshchiny', ed. I. A. Menitskii, *KA*, no. 67 (1934) pp. 28-92. The fourth issue was heavily censored by the central committee and the fifth was produced by the workers' group independently.

35   Among the members of this section were S.N. Prokopovich, P.P. Maslov, V.G. Groman and A.A. Manuilov. Manuilov, who served as chairman, became Minister of Education in the Provisional Government. See *Izv. MVPKa*, no. 15-16, Feb. 1916, pp. 103-4.

36   'K istorii Gvozdevshchiny', p. 35. The central committee asked the Special Council of Defence to provide the necessary funds and determine instances where they could be allocated, but no such relief was forth-coming.

37   Ibid. Also A. Kats, 'Bor'ba s bezrabotitsei v gody voiny', *Materialy po istorii professional'nogo dvizheniia v Rossii*, no. 3 (1925) pp. 79-83.

38   S.M. Schwarz, the secretary of the Moscow

workers' group, later considered the election of workers to the insurance boards as 'our great victory'. For his account of his six-month tenure as secretary, see Inter-University Project on Menshevism at Columbia University. 'Interview with Mensheviks' (Solomon Meerovich Shvarts), interview no. 2 (11 June 1962) pp. 1–29.

39   Like many of the groups' schemes, the workers' congress was not a new idea. In the aftermath of the 1905 revolution and particularly at the London (third) congress of the RSDRP, P.B. Akselrod had called for a non-party congress, initially as an alternative to the Duma and later to reassess the aims of the social democratic movement. Konovalov's remarks were made at a banquet during the Union of Towns' congress in March 1916. See *Burzhuaziia nakanune*, p. 95.

40   'Agent Blagoveshchenskii' (13 Jan. 1916), TsGAOR, f. 102 (1915), d. 347, 1. 214. Larin's participation is by no means certain. As late as 7 January he wrote from Stockholm that he disapproved of Levitskii's support of the workers' groups. This and an earlier letter in the same vein are in 'Iz istorii S-D internatsionalistskoi mysli vo vremia voiny', *Krasnaia letopis'*, no. 1 (10) (1924), pp. 179–81.

41   Polivanov's letter, dated 8 January 1916, is in TsGVIA, f. 369, op. 9, d. 6, 1. 53. For Khvostov's decision, see 'K istorii "rabochei gruppy" pri tsentral'nom voenno-promyshlennom komitete', ed. I.A. Menitskii, *KA*, no. 57 (1933) pp. 77–8.

42   'K istorii "rabochei gruppy"', p. 66.

43   G.G. Kasarov, 'Bor'ba proletariata protiv "rabochei gruppy" Moskovskogo Oblastnogo voenno-promyshlennogo komiteta', *Vestnik Moskovskogo universiteta, seriia ix: Istoriia*, no. 3 (1974) p. 43. See Bolshevik statement quoted in F. Romanov, 'Voenno-promyshlennye komitety i boikot ikh bol'shevikami', *Profsoiuzy SSSR*, no. 6 (1940) p. 54.

44   See Diakin, *Russkaia burzhuaziia i tsarizm*, p. 173.

45   *Trudy vtorogo s"ezda*, pp. 531–2, 628–9, 637,

646.

46 *Rech'*, no. 65, 7 Mar. 1916, p. 1. *Moskovskie Vedomosti* claimed that 'three-quarters of those present seemed to be representatives of the Jews, giving the hall an air of a Congress of Zionists and Revolutionaries', no. 49, 1 Mar. 1916, p. 1.

47 By March 1905 some 30 to 40 enterprises in Russia had elected factory elders. See A.F. Vovchik, *Politika tsarizma po rabochemu voprosu v predrevoliutsionnyi period, 1895-1904* (L'vov, 1964) p. 212.

48 *Den'*, no. 219, 11 Aug. 1916, p. 3.

49 *Izv. MVPKa*, no. 27-30, Aug.-Sept. 1916, pp. 192-3.

50 Ibid., no. 25-26, July 1916, p. 213; *Den'*, no. 240, 1 Sept. 1916, p. 4; no. 295, 26 Oct. 1916, p. 3; no. 356, 28 Dec. 1916, p. 5.

51 'K istorii "rabochei gruppy"', p. 59.

52 Gos. Duma, IV Sozyv: *Sten. otchety*, ses. 4, zas. 26 (22 Mar. 1916) cols 3750-2.

53 'K uchrezhdeniiu primiritel'nykh kamer', *Izv. TsVPKa*, no. 63, 26 Feb. 1916, p. 4.

54 'Zhurnal zasedaniia Soveta' (22 Mar. 1916), TsGIA, f. 150, op. 1, d. 64, l. 40.

55 *Izv. MVPKa*, no. 27-30, Aug.-Sept. 1916, pp. 194-7.

56 *Den'*, no. 268, 28 Sept. 1916, p. 5.

57 *Moskovskie Vedomosti*, no. 201, 31 Aug. 1916, p. 2.

58 Quoted in Shliapnikov, *Kanun semnadtsatogo goda*, p. 118.

59 Balabanov, *Ot 1905 k 1917*, p. 340.

60 'K istorii Gvozdevshchiny', pp. 46, 67. For strike figures per month based on factory and mine inspectors' tabulations, see Mints, 'Revoliutsionnaia bor'ba proletariata', pp. 24, 34.

61 *Novoe Vremia*, no. 14445, 25 May 1916, p. 3.

62 'Ruzskii Voennomu ministru' (2 Apr. 1916), TsGVIA, f. 369, op. 1, d. 205, l. 293; and 'Gvozdevshchina v dokumentakh', ed. M.K. Korbut, *Izvestiia Obshchestva arkheologii, istorii, i etnografii pri Kazanskom gosudarstvennom universitete*, no. 34 (1929) p. 267.

63 Kir'ianov, *Rabochie Iuga Rossii*, pp. 238-41, 254-7; 'K istorii Gvozdevshchiny', pp. 43-5, 59, and 'Gvozdevshchina v dokumentakh', pp.

221-3.

64   'K istorii Gvozdevshchiny', pp. 62, 82, and
      Izv. MVPKa, no. 21-2, May 1916, p. 215.
65   Dvinov, *Pervaia mirovaia voina i Ross.
      Sotsialdem.*, p. 131. Also, Tiutiukin, *Voina,
      mir revoliutsiia*, pp. 220-1.
66   Shliapnikov, *Kanun semnadtsatogo goda*, pp.
      103, 136, 293. Also, I.P. Leiberov,
      'Deiatel'nost' Petrogradskoi organizatsii
      bol'shevikov i ee vliianie na rabochee dvi-
      zhenie v Rossii v gody pervoi mirovoi voiny',
      in *Pervaia mirovaia voina*, p. 298.
67   *Izv. MVPKa*, no. 23-4, June 1916, p. 219, and
      'K istorii Gvozdevshchiny', p. 81.
68   *Russkoe Slovo*, no. 165, 17 July 1916, p. 5;
      no. 173, 27 July 1916, p. 4; no. 192, 20 Aug.
      1916, p. 3; no. 241, 19 Oct. 1916, p. 4.
      Schwarz was later called up to the armed
      forces. For a biographical sketch of Schwarz
      see L. Haimson, 'Preface' in S.M. Schwarz,
      The Russian Revolution of 1905, trans. G.
      Vakar (Chicago, 1967) pp. vii-xvii.
69   'K istorii Gvozdevshchiny', p. 82. The appeal
      was published in *Rech'*, no. 289, 20 Oct.
      1916, p. 4, and also appeared in *Izv. TsVPKa*,
      no. 161, 25 Oct. 1916, p. 2.
70   *Rech'*, no. 293, 24 Oct. 1916, p. 4.
71   *Den'*, no. 284, 15 Oct. 1916, p. 5; ibid., no.
      292, 23 Oct. 1916, p. 3.
72   'MVD Shuvaevu' (26 Nov. 1916), TsGVIA, f.
      369, op. 1, d. 206, 1. 438.
73   'Rezoliutsiia rabochei delegatsii soveshchan-
      iia oblastnykh voenno-promyshlennykh komite-
      tov' (13-15 Dec. 1916), TsGAOR, f. 102 (1917)
      d. 347, ch. 46, 1. 8. The assistance groups
      are referred to as 'oratorical collegia' in
      an undated police report in TsGAOR, f. 102
      (1917) d. 347, 11. 18-19.

CHAPTER 8:   THE CHALLENGE OF REVOLUTION

1   Gos. Duma, IV Sozyv: *Sten. otchety*, ses. 4,
     zas. 14 (25 Aug. 1915) col. 107.
2   *Burzhuaziia nakanune*, p. 90.
3   'Progressivnyi blok v 1915-1917 gg.', ed. N.
     Lapin, *KA*, no. 52 (1932) pp. 147-8.
4   Ibid., p. 149. This was the date of the

assassination of Tsar Paul I.

5   Diakin, *Russkaia burzhuaziia i tsarizm*, pp. 155-8; *Burzhuaziia nakanune*, pp. 65-71.
6   *Trudy vtorogo s"ezda*, p. 565.
7   Prokopovich and Nekrasov chaired the sessions devoted to food supply and labour. For the latter's interesting but not entirely accurate report on the congress, see 'Zhurnal soveshchaniia po ob"edineniiu deiatel'nosti predstavitelei soiuzov', *Izv. glav. kom. po snab. armii*, no. 17, 15 May 1916, p. 158.
8   *Trudy pervogo s"ezda predstavitelei metal. prom.*, pp. 26, 32-3, 39. Those critical of the VPKs included A.P. Meshcherskii, A.I. Putilov, M.A. Tokarskii and, revealing himself to be something of a fifth columnist in the Moscow committee, Jules Goujon.
9   *Pis'ma Imperatritsy Aleksandry Fedorovny k Imperatoru Nikolaiu II*, vol. II, p. 309.
10 *Izv. MVPKa*, no. 17-18, Mar. 1916, p. 2.
11 'Osobyi zhurnal Soveta ministrov' (15 Apr. 1916), TsGVIA, f. 369, op. 1, d. 206, l. 64.
12 'Osobyi zhurnal Soveta ministrov' (18 June 1916), ibid., d. 184, ll. 237-40 obv.
13 *Russkoe Slovo*, no. 95, 26 Apr. 1916, p. 3.
14 'Soveshchanie gubernatorov v 1916 g.', *KA*, no. 33 (1929) p. 158; *Izv. MVPKa*, no. 21-2, May 1916, p. 77.
15 *Burzhuaziia nakanune*, p. 106. Mrozovskii had in mind the Unions as well as the VPKs, though the ministry's plan concerned only the latter.
16 'Ruzskii Voennomu ministru' (2 Apr. 1916), TsGVIA, f. 369, op. 1, d. 205, l. 293.
17 Diakin, *Russkaia burzhuaziia i tsarizm*, p. 186. Shakhovskoi noted the committees' efforts to gain support from Allied ministers such as Albert Thomas, French Minister of Munitions.
18 See note 11, this chapter.
19 'Obzor o politicheskikh deiatel'nostiakh obshchestvennykh organizatsii s 1 marta po 15 aprelia 1916 g.', TsGAOR, f. 102 (1916) d. 338a, ll. 61-101. *Russkie Vedomosti*, no. 140, 18 June 1916, p. 3, reported the survey's existence. The Trudovik group consisted of moderate agrarian socialists loosely associated with the Socialist Revolutionary party.

20   *Russkoe Slovo*, no. 108, 11 May 1916, p. 3;
      ibid., no. 128, 4 June 1916, p. 1.
21   *Den'*, no. 128, 11 May 1916, p. 5. For the
      Samara case see *Izv. TsVPKa*, no. 178, 8 Dec.
      1916, p. 2.
22   'Pokhod protiv mobilizovannogo obshchestva',
      Izv. MVPKa, no. 25-6, July 1916, pp. 2-14.
23   'Min. Iustitsii, A.A. Khvostov, k prokura-
      toram sudnebykh palat' (8 July 1916), TsGVIA,
      f. 369, op. 1, d. 206, l. 164.
24   *Izv. glav. kom. po snab. armii*, no. 26-31, 1
      Dec. 1916, p. 230.
25   Diakin, *Russkaia burzhuaziia i tsarizm*, p.
      221. The procedures were drawn up by the
      Council of Ministers on 26 July and 12
      August.
26   *Izv. MVPKa*, no. 19-20, Apr. 1916, p. 130.
      Rech', no. 95, 6 Apr. 1916, p. 3 and no. 101,
      14 Apr. 1916, p. 3. The latter saw this as a
      'turn to the right for the bourgeoisie'.
27   'Doklady agentov "Blondinka" i "Zhurnalov"'
      (21 June 1916), TsGAOR, f. 102 (1916), d.
      343(3), ll. 211-12.
28   Ibid., d. 343(4), ll. 92-4.
29   *Trudy vtorogo s"ezda*, pp. 573-7.
30   *Izv. TsVPKa*, no. 107, 11 June 1916, p. 4;
      'Sluzhashchie i rabochie VPK-ov', *Izv. MVPKa*,
      no. 34 (Apr. 1917) pp. 26-7.
31   M.V. Novorusskii, 'Nachalo demobilizatsii',
      Izv. TsVPKa, no. 127, 2 Aug. 1916, p. 2.
32   *Burzhuaziia nakanune*, p. 140.
33   The British Library contains a copy of
      Borisov's plan, published under the title *The
      Russian Government's Plan of Future Railroad
      Construction* (New York, 1919).
34   *Trudy proiskhodivshego 20, 21, 22 okt. 1916
      g. soveshchaniia predstavitelei tsentral'nogo
      i oblastnykh VPK-ov dlia obsuzhdeniia plana
      zheleznodorozhnogo stroitel'stva na bli-
      zhaishee piatiletie* (Petrograd, 1916) pp. 2-
      8, 91-5. The total length of track as of Jan.
      1917 was 72,489 versts of which 6800 were
      occupied by the enemy (M.M. Shmukker, *Ocherki
      finansov i ekonomiki zheleznodorozhnogo
      transporta Rossii za 1913-1922 gody* (Moscow,
      1923) p. 71).
35   'Stenograficheskie otchety zasedaniia finan-
      sovo-ekonomicheskoi komissii obrazovannoi pri

SSPPiT' (31 Oct. 1916), TsGIA, f. 32, op. 1, d. 1852, l. 22.

36 Ibid. (8 Nov. 1916), ll. 62-3.

37 Ibid. (20 Feb. 1917), l. 123.

38 *Trudy...soveshchaniia...dlia plana zh.d. stroitel'stva*, p. 2.

39 The letter is in Semennikov, *Monarkhiia pered krusheniem*, pp. 279-82. An English translation of the relevant passages appears in Katkov, *Russia 1917*, pp. 256-7.

40 TsGAOR, f. 555, op. 1, d. 680, ll. 17-17 obv., contains a letter from N.I. Guchkov to Aleksandr, dated 2 November 1916.

41 Semennikov, *Monarkhiia pered krusheniem*, pp. 159-60 contains a report by Stuermer of 9 October 1916.

42 Katkov, *Russia 1917*, p. 260, views the letter and previous contacts between Guchkov and Alekseev as attempts by the former to win the support of the latter for 'a seditious movement'.

43 Guchkov in his memoirs claimed that the subject of a possible *coup* was broached for the first time at a meeting organized by M.M. Fedorov and attended by Guchkov, Konovalov, Prince L'vov and eight to ten others. The date of this meeting is in some doubt but in any case it did not occur before late September 1916. S.P. Mel'gunov, *Na putiakh k dvortsovomu perevorotu* (Paris, 1931) p. 146, also mentions this meeting as the starting point. More recently, V.I. Startsev, relying heavily on the diary of M.K. Lemke, a military correspondent at GHQ, has argued that as early as October-November 1915 Guchkov began to assemble a group which a year later would plot the *coup*. The evidence consists of visits by Guchkov and his VPK associates to various military officers including General Alekseev. Of course, it would be surprising had Guchkov conspired with total strangers. See Startsev, *Russkaia burzhuaziia i samoderzhavie*, pp. 189-90. The diary was published as M.K. Lemke, *250 dnei v tsarskoi stavke* (Petrograd, 1920).

44 Katkov, *Russia 1917*, p. 255 remarks that 'they [the generals] would not denounce them [the public organizations] to the Emperor or

to the state security authorities', but he has not seen the archives quoted here on p. 186.

45 Quoted in ibid., p. 257.

46 Guchkov was probably aware of Alekseev's plan for a military dictatorship proffered in June and the consternation this caused among court circles. For details see Manikovskii, *Boevoe snabzhenie*, vol. II, pp. 337–47.

47 See Diakin, *Russkaia burzhuaziia i tsarizm*, pp. 226–52; Pearson, *The Russian Moderates*, pp. 107–39.

48 *Burzhuaziia nakanune*, p. 141.

49 Gos. Duma, IV Sozyv: *Sten. otchety*, ses. 5, zas. 1, cols 35–50. Publication of the speech in the press was delayed for several weeks but mimeographed copies circulated widely in the country.

50 Diakin, *Russkaia burzhuaziia i tsarizm*, p. 258.

51 'Popytki organizovat' vserossiiskikh s"ezdov (9 Dec. 1916), TsGAOR, f. 102 (1916) d. 343(5), ll. 31–6. The account of these meetings in the police file has a tragicomic aspect. After being dispersed, delegates met in private apartments (L'vov's, Chelnokov's, Tret'iakov's) but were again found and dispersed. A veritable chase around the city ended in the Prague restaurant where delegates promised to eat and not to discuss politics. The resolutions are in *Burzhuaziia nakanune*, pp. 155, 158.

52 TsGAOR, f. 102 (1916) d. 343(4), l. 261.

53 *Izv. TsVPKa*, no. 215, 13 Apr. 1917, pp. 2–3.

54 See note 42.

55 V.A. Maklakov and V.V. Shul'gin were aware of the existence of the plot to assassinate Rasputin. According to Diakin, *Russkaia burzhuaziia i tsarizm*, p. 263, Miliukov 'and possibly a number of other Kadets' were also informed.

56 The most detailed work on these plots has been done by Mel'gunov, *Na putiakh k dvortsovomu perevorotu*. This can be supplemented by testimony given to the Extraordinary Investigation Commission of the Provisional Government published as *Padenie tsarskogo rezhima*; G. Aronson, *Rossiia nakanune revol-*

iutsii (New York, 1962) pp. 109–24; Katkov, *Russia 1917*, pp. 244–8, 296–300; and Diakin, *Russkaia burzhuaziia i tsarizm*, pp. 298–311.

57 Buryshkin, *Moskva kupecheskaia*, p. 316.

58 Countess Panina Collection, Columbia University Russian Archive, papka 2, folder 10, letter from Astrov to Mel'gunov, 10 Jan. 1931.

59 Laverychev, *Po tu storonu barrikad*, p. 159, casts his net wide and comes up with Riabushinskii as a member of L'vov's group, apparently mistaking him for N.M. Kishkin. See Diakin, *Russkaia burzhuaziia i tsarizm*, pp. 300–3. The importance of the conspiratorial ties lay in the selection of ministers for the Provisional Government. See Katkov, *Russia 1917*, pp. 501–2.

60 *Poslednie novosti*, no. 5651, 13 Sept. 1936, p. 2.

61 Baron A.F. Meyendorff Papers, Valtionarkisto (Finnish State Archives), Helsinki, correspondence with A.I. Guchkov, 8 Feb. 1931, Kansio 2.

62 'Doklad okhrannogo otdeleniia v Moskve' (n.d.), TsGAOR, f. 102 (1917) d. 347, l. 25.

63 Protopopov's letter is in TsGVIA, f. 369, op. 1, d. 206, l. 437. The report of 2 January is in 'V ianvar' i fevral' 1917 g., "Iz donesenii sekretnykh agentov A.D. Protopopova"', *Byloe*, no. 13 (1918) p. 94.

64 Of the three, Ia.I. Anasovskii and Ia.S. Ostapenko were not in Petrograd and the third was Abrosimov.

65 A.G. Shliapnikov, *Semnadtsatyi god*, 4 vols, (Moscow, 1923–31) vol. I, pp. 224–5.

66 Ev. Maevskii, *Kanun revoliutsii* (Petrograd, 1918) p. 10.

67 The police report of the meeting is in *Burzhuaziia nakanune*, pp. 180–3. Of the 35 present only four, Pereverzev, Zernov, Bublikov and the left Kadet, Adzhemov, spoke in favour of continuing the work of the workers' group or reconstituting it. For Miliukov's letter see *Rech'*, no. 39, 10 Feb. 1917, p. 3.

68 See interview in *Rech'*, no. 43, 14 Feb. 1917, p. 2.

69 See Gos. Duma, IV Sozyv: *Sten. otchety*, ses.

5, zas. 21 (12 Feb. 1917) cols 1516–32.

70   See, for example, Gaponenko, *Rabochii klass Rossii v 1917*, pp. 138–9, and Tiutiukin, *Voina, mir, revoliutsiia*, p. 298.

71   Pearson, *The Russian Moderates*, p. 138.

72   *Izv. MVPKa*, no. 34 (Apr. 1917) pp. 16–23; Diakin, *Russkaia burzhuaziia i tsarizm*, p. 337.

73   S. Dmitrievskii, 'Vse dlia pobedy', *Izv. TsVPKa*, no. 208, 13 Mar. 1917, p. 2.

74   'Revoliutsiia i oblastnye i mestnye VPKy', ibid., pp. 5–6.

75   *Izv. MVPKa*, no. 34 (Apr. 1917) pp. 25, 28–34; Laverychev, *Monopol. kapital v tekst. prom.*, p. 392.

76   *Izv. TsVPKa*, no. 208, 13 Mar. 1917, pp. 2–4.

77   Shliapnikov, *Semnadtsatyi god*, vol. II, pp. 236–8.

78   For rather sketchy accounts of these meetings see N. Sukhanov, *The Russian Revolution, 1917: A Personal Record*, ed. and trans. J. Carmichael (London, 1955) pp. 14–15; V. Chernov, *The Great Russian Revolution*, trans. P. Mosely (New York, 1966) pp. 101–2; I.P. Leiberov, 'Vtoroi den' Fevral'skoi revoliutsii', in *Sverzhenie samoderzhaviia: Sbornik statei* (Moscow, 1970) p. 118; Katkov, *Russia 1917*, p. 476.

79   Sukhanov, *The Russian Revolution, 1917*, p. 59. Oskar Anweiler, *The Soviets: The Russian Workers, Peasants, and Soldiers Councils, 1905–1921*, trans. R. Hein (New York, 1974) p. 104. Sukhanov describes Bogdanov as 'the most active member of the Ex. Com.'. Anweiler goes so far as to say that the 'decisive step in forming the Petrograd soviet was taken by members of the central Workers' Group who were released from prison on 27 February'.

80   E.N. Burdzhalov, *Vtoraia russkaia revoliutsiia: Moskva, front, periferiia* (Moscow, 1971) pp. 212, 215. Also, *Revoliutsionnoe dvizhenie v Rossii posle sverzheniia samoderzhaviia* (Moscow, 1957) p. 241; A.Ia. Grunt, *Moskva 1917: Revoliutsiia i kontrrevoliutsiia* (Moscow, 1976) p. 68; and Donald J. Raleigh, 'Revolutionary Politics in Provincial Russia: the Tsaritsyn "Republic" in 1917', *Slavic Review*, vol. XL (1981) pp. 197–

8. True, in the case of Tsaritsyn, six members of the workers' group sat in both organs of dual power.

81 See *Ekonomicheskoe polozhenie Rossii nakanune ...revoliutsii*, vol. I, pp. 165-8, 205-8.

82 GBIL, f. 260, papka 2, ed.kh. 20, ll. 13-14, 16-17, 18. Citations are from the original version. For more on the All-Russian Union of Trade and Industry see V.Ia. Laverychev, 'Vserossiiskii soiuz torgovli i promyshlennosti', *IZ*, vol. LXX (1961) pp. 35-55.

83 'Nakaz o poriadke obrazovaniia i deistvii VPKa' and 'Rezoliutsiia o polozhenii VPK-ov', TsGAGM, f. 1082, op. 1, d. 1, ll. 17-19. Another 45 were to consist of district committee representatives and specially invited people.

84 'Demokratizatsiia VPK-ov', *Izv. MVPKa*, no. 35, June 1917, p. 1.

85 Razumovskaia, 'Tsentral'nyi voenno-promyshlennyi komitet', p. 351, mentions that 54 Mensheviks and 15 Socialist Revolutionaries were present but does not indicate if there were other workers' delegates.

86 *Izv. TsVPKa*, no. 230, 25 May 1917, p. 2, gives the Presidium as follows: S.A. Smirnov, S.N. Tret'iakov, S.S. Smirnov, Ia.A. Galiashkin, N.N. Kutler, P.P. Kozakevich, A.A. Bublikov, N.F. Fon-Ditmar, D.V. Sirotkin, N.G. Raiskii and S.F. Zavalishin.

87 *Utro Rossii*, no. 120, 16 May 1917, p. 1.

88 Ibid., no. 121, 17 May 1917, p. 3.

89 *Gornozavodskoe delo*, no. 26-7, 1 July 1971, p. 16071.

90 Ibid., p. 16073; *Torgovo-promyshlennaia gazeta*, no. 103, 19 May 1917, pp. 2-3.

91 *Utro Rossii*, no. 124, 20 May 1917, p. 5.

92 *Izvestiia Zemgora*, no. 7, 27 May 1917, p. 5.

93 N. Volkovskii, 'Vpechatlenie', *Utro Rossii*, no. 123, 19 May 1917, p. 3; *Torgovo-promyshlennaia gazeta*, no. 105, 21 May 1917, pp. 2-3; ibid., no. 106, 24 May 1917 p. 3.

94 *Izv. MVPKa*, no. 36 (July 1917) p. 13; ibid., no. 45-6 (Oct. 1917) p. 16.

95 *Ekonomicheskoe polozhenie Rossii nakanune... revoliutsii*, vol. I, pp. 264-7, 596.

96 Ibid., pp. 361-2; 'O vydache dopolnitel'nogo avansa MVPKu po priniatym komitetom zakazam

GIU na obuv' (21 Aug. 1917), TsGVIA, f. 369, op. 1, d. 402, l. 41.

97 These factors induced the Vankov Organization to attempt a takeover of factories previously contracting with the VPKs (and the Zemgor), quite the reverse of what the Moscow committee had suggested earlier. See *Izvestiia Zemgora*, no. 5, 13 May 1917, p. 11; N.A. Ivanova, 'Prinuditel'nye ob"edineniia v Rossii v gody pervoi mirovoi voiny', in *Ob osobennostiakh imperializma v Rossii*, pp. 247–8.

98 Zagorsky, *State Control of Industry*, p. 229; Laverychev, *Monopol. kapital v tekst. prom.*, p. 395.

99 *Izv. TsVPKa*, no. 209, 18 Mar. 1917, p. 4; 'Tsentropotash', ibid., no. 270, 19 Oct. 1917, p. 3.

100 *Izv. MVPKa*, no. 40, Aug. 1917, p. 12.

101 Ibid., no. 35, June 1917, p. 13; no. 37, July 1917, p. 12.

102 *Ekonomicheskoe polozhenie Rossii nakanune... revoliutsii*, vol. I, pp. 558–60.

103 *Izv. MVPKa*, no. 43–4, Oct. 1917, p. 6.

104 'Vo vremennoe pravitel'stvo' (Sept. 1917), TsGAGM, f. 1082, op. 1, d. 1, l. 32.

105 'Pravila o poriadke likvidatsii deiatel'nosti oblastnykh i mestnykh VPK-ov' (26 Sept. 1917), ibid., ll. 34–5.

106 *Izv. MVPKa*, no. 36, July 1917, p. 8.

107 'O peremeshchenii promyshlennykh tsentrov Rossii', *Proizvoditel'nye sily Rossii*, no. 12, Aug. 1917, pp. 3–4.

108 'Pochin i primer', *Izv. MVPKa*, no. 45–6, Oct. 1917, p. 1.

109 *Birzhevye Vedomosti*, no. 16472, 6 Oct. 1917, p. 4; see M.Ia. Lapirov-Skoblo, *Rabota nauchno-tekhnicheskikh uchrezhdenii respubliki, 1918–1919* (Moscow, 1919) pp. 33, 50, 203.

110 'Proekt rezoliutsii TsVPKa "po voprosam tekushchego momenta"' (22 July 1917), TsGVIA, f. 13251, op. 11, d. 7, ll. 12–16.

111 'Grazhdane-rabochie', TsGVIA, f. 13251, op. 11, d. 7, l. 16. There is no date attached though the phrase 'eight months have passed' would put it towards the end of October.

112 'My dolzhny rabotat'', *Izv. MVPKa*, no. 47,

Nov. 1917, p. 1.

113 *Proizvoditel'nye sily Rossii*, no. 17-20, Oct. 1917, pp. 1-2, 18-19, 20. Much the same prospect was embodied in the sixteen points concerning the demobilization of industry which V.I. Grinevetskii presented to the Moscow congress of the Union of Engineers on 4-6 January 1918. See *Biulleten Moskovskogo oblastnogo biuro i Moskovskogo otdeleniia Vserossiiskogo soiuza Inzhenerov* (Moscow, 1918) pp. 33-7. For his more general, long-term prognosis see *Poslevoennye perspektivy russkoi promyshlennosti* (Moscow, 1919).

NOTES TO THE EPILOGUE

1 *Izv. TsVPKa*, no. 276, 9 Dec. 1917, p. 1; *Izv. MVPKa*, no. 51, Jan. 1918, p. 16.

2 *Sbornik dekretov i postanovlenii po narodnomu khoziaistvu* (Moscow, 1918) p. 150.

3 Ibid.

4 *Izvestiia Moskovskogo Narodno-promyshlennogo komiteta*, no. 52-3 (Apr. 1918) p. 11.

5 *Sbornik dekretov...po narodnomu khoziaistvu*, p. 150.

6 *Izv. MVPKa*, no. 54-5 (May 1918) p. 3; *Izvestiia tsentral'nogo narodno-promyshlennogo komiteta*, no. 292, 5 Apr. 1918, p. 2.

7 *Sbornik dekretov...po narodnomu khoziaistvu*, p. 149.

8 M.S. Margulies, *God interventsii, sent. 1918 - aprel' 1919 g.* (Berlin, 1923) p. 214.

Select Bibliography

Some sources with only one citation in the notes have not been included in the Bibliography.

MANUSCRIPT AND ARCHIVAL SOURCES

1 *Sources outside the Soviet Union*

Columbia University, Bakhmet'ev Archive (formerly the Archive of Russian and East European History and Culture), New York: Countess Panina Collection

Columbia University, Inter-University Project on Menshevism at Columbia University, interviews with Mensheviks: Solomon Meerovich Shvarts (1962)

National Archives Division, Washington, D.C.: Record Group 261, Russian Government Supply Committee in the United States

Public Record Office, London: FO 371, general correspondence of the Foreign Office

School of Slavonic and East European Studies, London: Sir Bernard Pares Papers

Valtionarkisto (Finnish State Archives), Helsinki: Baron Aleksandr Feliksovich Meyendorff Papers

2 *Sources in the Soviet Union*

Central State Archive of the City of Moscow (TsGAGM), Moscow: f. 1082 - Moskovskii voenno-promyshlennyi komitet (Moscow War-Industries Committee)

Central State Archive of the October Revolution (TsGAOR), Moscow: f. 102 - Departament politsii (Department of Police), 1915, 1916, 1917; f.555

- Lichnyi fond A.I. Guchkova (Personal Archive
of A.I. Guchkov)
Central State Historical Archive (TsGIA),
Leningrad: f. 32 - Sovet s"ezdov promyshlennosti
i torgovli (Association of Industry and Trade);
f. 150 - Petrogradskoe Obshchestvo fabrikantov i
zavodchikov (Petrograd Society of Factory and
Mill Owners); f. 1131 - Komitet dlia zavedy-
vaniia snabzhenii syrem khlopchatobumazhnykh
fabrik (Committee for the Supply of Factories
with Raw Cotton); f.1276 - Sovet ministrov -
Kantseliariia (Council of Ministers - Chan-
cellery)
Central State Military History Archive (TsGVIA),
Moscow: f. 369 - Osoboe soveshchanie po oborone
(Special Council of Defence); f. 12642 -
Vserossiiskie Zemskii i Gorodskoi soiuzy (All-
Russian Zemstvo and Town Unions); f. 13251 -
Tsentral'nyi voenno-promyshlennyi komitet
(Central War-Industries Committee)
Lenin Library - Manuscript Division (GBIL - Otdel
rukopisei), Moscow: f. 225 - Fond A.V.
Peshekhonova (A.V. Peshekhonov Archive); f. 260
- Fond P.P. Riabushinskogo (P.P. Riabushinskii
Archive)

3 *Unpublished Manuscripts*

Gisin, S.L., 'Vserossiiskii Zemskii soiuz:
politicheskaia evoliutsiia s iiul' 1914 po fev.
1917', candidate's dissertation, 1947, deposited
in GBIL.
Gleason, Wm E., 'The All-Russian Union of Towns
and the All-Russian Union of Zemstvos in World
War I: 1914-1917', Ph.D. dissertation, Indiana
University, 1972.
Godfrey, J.F., 'Bureaucracy, Industry and Politics
in France during the First World War: a Study of
Some Interrelationships', submitted for D.Phil.
thesis, Oxford University, 1971.
Goldberg, C., 'The Association of Industry and
Trade, 1906-17: the Successes and Failures of
Russia's Organized Businessmen', Ph.D.
dissertation, University of Michigan, 1974.
Goldstein, E.R., 'Military Aspects of Russian
Industrialization', Ph.D. dissertation, Case
Western Reserve, Cleveland, Ohio, 1971.
Korelin, A.P., 'Monopolii v metalloobraba-

tyvaiushchei promyshlennosti Rossii i ikh
antirabochaia politika v gody pervoi mirovoi
voiny', candidate's dissertation, Moscow State
University, 1964, deposited in GBIL.

Menashe, L., 'Alexander Guchkov and the Origins of
the Octobrist Party', Ph.D. dissertation, New
York University, 1966.

Miftiev, G.K., 'Artilleriiskaia promyshlennost'
Rossii v period pervoi mirovoi voiny',
candidate's dissertation, Leningrad State
University, 1953, deposited in GBIL.

Razumovskaia, N.I., 'Tsentral'nyi voenno-
promyshlennyi komitet', candidate's dissertation, Moscow State University, 1947, deposited
in GBIL.

Roosa, R.A., 'The Association of Industry and
Trade 1906-1914: an Examination of the Economic
Views of Organized Industrialists in Prerevol-
utionary Russia', Ph.D. dissertation, Columbia
University, 1967.

Ruckman, J.A., 'The Business Elite of Moscow: a
Social Enquiry', Ph.D. dissertation, Northern
Illinois University, 1975.

Voronkova, S.V., 'Materialy Osobogo soveshchaniia
po oborone gosudarstva', candidate's dissertation, Moscow State University, 1971, deposited
in GBIL.

West, J.L., 'The Moscow Progressists: Russian
Industrialists in Liberal Politics, 1905-1914',
Ph.D. dissertation, Princeton University, 1975.

PRINTED SOURCES

1 *Primary Sources: Pamphlets, Collections of
  Documents and Speeches, Official Publications,
  Memoirs*

An asterisk indicates a publication of the War-
Industries Committees.

Aktsionerno-paevye obshchestva v Rossii (Moscow,
  1913).

'Antivoennaia rabota bol'shevikov v gody pervoi
  mirovoi voiny, 1914-1917 gg., *Istoricheskii
  arkhiv*, no. 5 (1961) pp. 74-107.

Bart, A., 'Na fronte artilleriiskogo snabzheniia',
  Byloe, vol. V (XXXIII) (1925) pp. 188-219.

Bazanov, Ia., 'Sem' mesiatsev partiinoi raboty v Khar'kove', *Proletarskaia revoliutsiia*, no. 9 (1922) pp. 77-137.

Bol'sheviki v gody imperialisticheskoi voiny, 1914-fev. 1917 (Leningrad, 1939).

Buryshkin, P.A., *Moskva kupecheskaia* (New York, 1954).

Burzhuaziia nakanune fevral'skoi revoliutsii, ed. B.B. Grave (Moscow and Leningrad, 1927).

Deiatel'nost' Moskovskogo oblastnogo voenno-promyshlennogo komiteta i ego otdelov po 31 ianvaria 1916 g. (Moscow, 1917). *

Deiatel'nost' Moskovskogo voenno-promyshlennogo komiteta i ego otdelov na I/VI 1917 (Moscow, 1917). *

Deiatel'nost' oblastnykh i mestnykh voenno-promyshlennykh komitetov na 10 fevralia 1916 goda, 3 vols (Petrograd, 1916). *

Deviaty i ocherednoi S"ezd predstavitelei promyshlennosti i torgovli, *Doklad Soveta S"ezdov o merakh k razvitiiu proizvoditel'nykh sil Rossii* (Petrograd, 1915).

Documents of Russian History, 1914-1917, ed. Frank A. Golder (Gloucester, Mass., 1964).

Doklad schetnogo otdela, 3 vols (Petrograd, 1916). *

Dokumenty Biuro TsK R.S.D.R.P. v Rossii (iiul' 1914-fevral' 1917 gg.), *Voprosy istorii KPSS*, no. 8 (1965) pp. 90-100.

Ekonomicheskoe polozhenie Rossii nakanune Velikoi Oktiabr'skoi sotsialisticheskoi revoliutsii, 3 vols (Moscow and Leningrad, 1957-67) vols 1-2.

Finansovyi otchet s nachala deiatel'nosti Moskovskogo voenno-promyshlennogo komiteta po 1 avgusta 1916 (Moscow, 1917). *

Gorbachev, I.A., *Khoziaistvo i finansy voenno-promyshlennykh komitetov* (Moscow, 1919).

Gosudarstvennaia Duma, *Stenograficheskie otchety*, III Sozyv, sessii 1, 5 (St Petersburg, 1908, 1912); IV Sozyv, sessii 1, 4, 5 (St Petersburg, 1913; Petrograd, 1915-17).

Gosudarstvennyi Sovet, *Stenograficheskie otchety*, sessiia 11 (Petrograd, 1915).

Guchkov, A.I., *A.I. Guchkov v tret'ei Gosudarstvennoi dume, 1907-1912 gg.* (St Petersburg, 1912).

Guchkov, A.I., *Doklad po zakonoproektu ob otpuske na 1912 goda sredstv na popolnenie zapasov i materialov* (St Petersburg, 1912).

Guchkov, A.I., *Rechi po voprosam gosudarstvennoi oborony i ob obshchei politike, 1908-17* (Petrograd, 1917).

'Gvozdevshchina v dokumentakh', ed. M.K. Korbut, *Izvestiia Obshchestva arkheologii, istorii, i etnografii pri Kazanskom gosudarstvennom universitete*, no. 34 (Kazan', 1929) pp. 220-70.

Iakhontov, A.N. (ed.), 'Tiazhelye dni: sekretnye zasedaniia Soveta ministrov 16 iiulia-2 sentiabria 1915 goda', *Arkhiv russkoi revoliutsii*, vol. XVIII (Berlin, 1926) pp. 5-136.

Ipat'ev, V.N., *Zhizn' odnogo khimika, Vospominaniia*, 2 vols (New York, 1945) vol. I.

Istoriia organizatsii upolnomochennogo po zagotovleniiu snariadov po frantsuzskomu obraztsu general-maiora S.N. Vankova, 1915-1918 (Moscow, 1918).

'Iz dnevnika Sukhomlinova', *Dela i dni*, no. 1 (1920) pp. 220-37.

'Iz istorii S-D internatsionalistskoi mysli vo vremia voiny', *Krasnaia letopis'*, no. 1(10) (1924) pp. 116-82.

'Iz vospominaniia A.I. Guchkova', *Poslednie novosti* (Paris, Aug.-Sept. 1936).

'Iz zapisnoi knizhki arkhivista', *Krasnyi arkhiv*, no. 59 (1933) pp. 143-8.

'K istorii Gvozdevshchiny', ed. I.A. Menitskii, *Krasnyi arkhiv*, no. 67 (1934) pp. 28-92.

'K istorii kontserna br. Riabushinskikh', ed. I.F. Gindin, in *Materialy po istorii SSSR*, vol. 6 (Moscow, 1959) pp. 603-40.

'K istorii "rabochei gruppy" pri tsentral'nom voenno-promyshlennom komitete', ed. I.A. Menitskii, *Krasnyi arkhiv*, no. 57 (1933) pp. 43-82.

'Kadety v dni Galitsiiskogo razgroma', ed. N. Lapin, *Krasnyi arkhiv*, no. 59 (1933) pp. 110-44.

'Kafengauz, L.B., 'Snabzhenie strany mineral'nym toplivom vo vremia voiny', *Trudy komissii po izucheniiu sovremennoi dorogovizny*, 4 vols (Moscow, 1915) vol. 2, pp. 231-77.

Khoziaistvennaia zhizn' i ekonomicheskoe polozhenie naseleniia Rossii za pervoi deviat' mesiatsev voiny (Petrograd, 1916).

Kievskii oblastnyi voenno-promyshlennyi komitet (Kiev, 1916). *

Krestovnikov, N.K., *Semeinaia khronika Kres-*

tovnikovykh, 3 vols (Moscow, 1903-4).

'Le plan de Mihajl Rjabusinskij: un projet de concentration industrielle en 1916', ed. M.L. Lavigne, *Cahiers du monde russe et soviétique*, no. 1 (1964) pp. 90-104.

Lenin, V.I., *Polnoe sobranie sochinenii*, 5th edn, 55 vols (Moscow, 1958-65) vols 31, 47, 49.

Lenin, V.I., *Sochineniia*, 4th edn, 35 vols (Moscow, 1941-50) vol. 19.

Lichnyi sostav voenno-promyshlennykh komitetov (Petrograd, 1915). *

Lloyd George, D., *War Memoirs*, 6 vols (London, 1933-8) vols 1-2.

Lur'e, E.S., *Organizatsiia organizatsii torgovo-promyshlennykh interesov v Rossii* (St Petersburg, 1913).

Maevskii, Ev., *Kanun revoliutsii* (Petrograd, 1918).

Manikovskii, A.A., *Boevoe snabzhenie russkoi armii v mirovuiu voinu*, 2nd edn, 2 vols (Moscow, 1930-2).

Margulies, M.S., *God interventsii, sent. 1918-aprel' 1919 g.* (Berlin, 1923).

Materialy k uchetu rabochego sostava i rabochego rynka, 2 vols (Petrograd, 1916-17). *

Mendeleev, D.I., *Problemy ekonomicheskogo razvitiia Rossii* (Moscow, 1960).

Miliukov, P.N., *Political Memoirs, 1905-1917* (Ann Arbor, Mich., 1967).

Miliukov, P.N., *Rossiia na perelome*, 2 vols (Paris, 1927) vol. 1.

Monopolii v metallurgicheskoi promyshlennosti Rossii, 1900-1917, Dokumenty i materialy (Moscow and Leningrad, 1963).

Monopolisticheskii kapital v neftianoi promyshlennosti Rossii, 1883-1914, and 1914-1917, Dokumenty i materialy, 2 vols (Moscow and Leningrad, 1961-73).

Narodnoe khoziaistvo v 1913 godu (Petrograd, 1914).

Narodnoe khoziaistvo v 1915 godu (Petrograd, 1918).

Ob organizatsii osoboi komissii po promyshlennosti pri I.R.T. Obshchestve i ob ee zadachakh (Petrograd, 1914).

Organizatsiia voenno-promyshlennykh komitetov (Petrograd, 1915). *

Organizatsiia raspredeleniia voennykh zakazov po

raionam Imperii (Petrograd, 1915). *

Osobye soveshchaniia i komitety voennogo vremeni (Petrograd, 1917).

Otchet o deiatel'nosti Osobogo soveshchaniia i ob"edineniia meropriiatii po perevozke za sentiabr' 1915-sentiabr' 1916 g. (Petrograd 1916).

Otchet o deiatel'nosti otdelov TsVPKa na 1 noiabria 1915 g. (Petrograd, 1916). *

Otchety otdelov TsVPKa k tret'emu Vserossiiskomu s"ezdu voenno-promyshlennykh komitetov v Moskve (Petrograd, 1917). *

Outline of Activities of the Central War Industrial Committee of Russia, An (New York, 1918). *

Padenie tsarskogo rezhima. Stenograficheskie otchety voprosov i pokazanii danny v 1917 g., v Chrezvychainoi sledstvennoi komissii vremennogo pravitel'stva, ed. P.E. Shchegolev, 7 vols (Moscow and Leningrad, 1924-7).

Paleologue, Maurice, *An Ambassador's Memoirs*, trans. by F.A. Holt, 3 vols (London, 1923-5) vols 1-2.

Panov, N.F., 'Moia rabota v Samare', *Krasnaia byl'*, no. 1 (1922) pp. 3-49.

Pares, B., *Day by Day with the Russian Army* (London, 1915).

Pares, B., *My Russian Memoirs* (London, 1931).

'Perepiska V.A. Sukhomlinova s N.N. Ianushkevichem', *Krasnyi arkhiv*, no. 1 (1922) pp. 209-62.

Pis'ma Imperatritsy Aleksandry Fedorovny k Imperatoru Nikolaiu II, ed. V.D. Nabokov, 2 vols (Berlin, 1922).

Polivanov, A.A., *Iz dnevnikov i vospominanii po dolzhnosti voennogo ministra i ego pomoshchnika* (Moscow, 1924).

Predstavitel'stvo obshchestvennykh grupp v voenno-promyshlennykh komitetakh (Petrograd, 1916). *

'Produgol'' (k voprosu o finansovom kapitale v Rossii), *Krasnyi arkhiv*, no. 18 (1926) pp. 119-48.

'Progressivnyi blok v 1915-17 gg.', ed. N. Lapin, *Krasnyi arkhiv*, no. 50-1 (1932) pp. 117-60; no. 52 (1933) pp. 143-96; no. 56 (1933) pp. 80-135.

Prologue to Revolution: Notes of A.N. Iakhontov on the Secret Meetings of the Council of Ministers, 1915, ed. Michael Cherniavsky (Englewood Cliffs, N.J., 1967).

Promyshlennost' i torgovlia v zakonodatel'nykh uchrezhdeniiakh, 1909-1912 gg. (St Petersburg, 1913).

Rabochee dvizhenie v Petrograde v 1912-1917 gg., Dokumenty i materialy (Leningrad, 1958).

Revoliutsionnoe dvizhenie v Rossii posle sverzheniia samoderzhaviia (Moscow, 1957).

Rodzianko, M.V., 'Krushenie imperii', *Arkhiv russkoi revoliutsii*, vol. XVII (Berlin, 1926) pp. 5-169.

Sbornik dekretov i postanovlenii po narodnomu khoziaistvu (Moscow, 1918).

Shakhovskoi, V.N., *Sic Transit Gloria Mundi* (Paris, 1952).

Shliapnikov, A.G., *Kanun semnadtsatogo goda: Vospominaniia i dokumenty o rabochem dvizhenii i revoliutsionnom podpol'e za 1914-1916 gg.*, 2nd edn (Moscow, 1923).

Shliapnikov, A.G., *Semnadtsatyi god*, 4 vols (Moscow, 1923-31) vols 1 and 2.

Sobranie uzakonenii i rasporiazhenii Pravitel'stva (Petrograd, 1915-16).

Soobshchenie predstavitelei 34 voenno-promyshlennykh komitetov, 7-31 dekabria 1915 g. (Petrograd, 1916). *

Spisok fabrik i zavodov g. Moskvy i Moskovskoi gubernii sostavlen fabrichnymi inspektorami Moskovskoi gubernii po dannym 1916 goda (Moscow, 1916).

Spisok sobstvennykh predpriiatii voenno-promyshlennykh komitetov (Petrograd, 1917). *

Spisok voenno-promyshlennykh komitetov (Petrograd, 1916). *

Statisticheskaia razrabotka dannykh o deiatel'nosti glavnogo po snabzheniiu armii komiteta Vserossiiskogo Zemskogo Soiuza i Vserossiiskogo Soiuza Gorodov i voenno-promyshlennykh komitetov po vypolneniiu planovykh zakazov voennogo vedomstva pervoi ocheredi (Moscow, 1917).

Sukhanov, N., *The Russian Revolution, 1917: A Personal Record*, ed. and trans. J. Carmichael (London, 1955).

Sukhomlinov, V.A., *Vospominaniia* (Berlin, 1924).

Trudy ekonomicheskogo soveshchaniia, 3-4 ianvaria, 1916 g. (Moscow, 1916).

Trudy pervogo s"ezda predstavitelei metalloobrabatyvaiushchei promyshlennosti, 29 fev.-1 marta 1916 g. (Petrograd, 1916).

Trudy proiskhodivshego 20, 21, 22 okt. 1916 soveshchaniia predstavitelei tsentral'nogo i oblastnykh voenno-promyshlennykh komitetov dlia obsuzhdeniia plana zheleznodorozhnogo stroitel'-stva na blizhaishee piatiletie (Petrograd, 1916). *

Trudy s"ezda predstavitelei voenno-promyshlennykh komitetov, 25-27 iiulia 1915 g. (Petrograd, 1915). *

Trudy vtorogo s"ezda predstavitelei voenno-promyshlennykh komitetov, 26-29 fevralia 1916 g. (Petrograd, 1916). *

Tsentral'noe statisticheskoe upravlenie, SSSR, *Rossiia v mirovoi voine* (Moscow, 1925).

'V ianvar' i fevral' 1917 g., "Iz donesenii sekretnykh agentov A.D. Protopopova"', *Byloe*, no. 13 (1918) pp. 91-123.

Velikaia Rossiia, Sbornik statei po voennym i obshchestvennym voprosam, ed. V.P. Riabushin-skii, 2 vols (Moscow, 1910-11).

Zalezhskii, V.N., *Iz vospominanii podpol'shchika* (Khar'kov, 1931).

Zhurnal zasedanii vos'mogo ocherednogo s"ezda predstavitelei promyshlennosti i torgovli, sostoiavshchegosia 2, 3, i 4 maia 1914 g., v Petrograde (Petrograd, 1915).

Zhurnaly Osobogo soveshchaniia dlia obsuzhdeniia i ob"edineniia meropriiatii po oborone gosudarstva (Osoboe soveshchanie po oborone gosudarstva), 1915-1918 gg., 2 parts (Moscow, 1975).

2 Newspapers and Periodicals

Birzhevye Vedomosti (Petrograd, 1914-17).

Den' (Petrograd, 1915-17).

Gornozavodskoe delo (Khar'kov, 1915-17).

Izvestiia glavnogo komiteta po snabzheniiu armii; in 1917 *Izvestiia Zemgora* (Moscow, 1915-17).

Izvestiia Moskovskogo voenno-promyshlennogo komiteta (Moscow, 1915-18). *

Izvestiia Obshchestva zavodchikov i fabrikantov Moskovskogo promyshlennogo raiona (Moscow, 1914-15).

Izvestiia Osobogo soveshchaniia po toplivu (Petrograd, 1917).

Izvestiia tsentral'nogo voenno-promyshlennogo komiteta (Petrograd, 1915-18). *

Kommercheskii telegraf (Moscow, 1915).
Moskovskie Vedomosti (Moscow, 1915-16).
Neftianoe delo (Baku, 1914).
Novoe Vremia (Moscow, 1914-16).
Proizvoditel'nye sily Rossii (Moscow, 1916-17). *
Promyshlennaia Rossiia (Moscow, 1915).
Promyshlennost' i Torgovlia (Petrograd, 1908-17).
Rech' (Petrograd, 1915-17).
Russkie Vedomosti (Moscow, 1915-16).
Russkoe Slovo (Moscow, 1915-17).
Torgovo-promyshlennaia gazeta (Petrograd, 1917).
Trudy komissii po promyshlennosti v sviazi s voinoi (Petrograd, 1915).
Utro Rossii (Moscow, 1915-17).
Zhurnaly tsentral'nogo voenno-promyshlennogo komiteta (Petrograd, 1915-16).

3 *Secondary Sources*

Abramov, A.A., 'Bor'ba bol'shevistskikh organizatsii protiv sotsial-shovinizma i tsentrizma v Rossii (1914-fev. 1917)', *Voprosy istorii KPSS*, no. 11 (1963) pp. 44-56.

Adamov, V.V., 'Iz istorii mestnykh voenno-ekonomicheskikh organizatsii v gody pervoi mirovoi voiny (voenno-promyshlennye komitety na Urala)' in *Voprosy istorii Urala* (Sverdlovsk, 1958) pp. 82-95.

Anderson, Perry, *Lineages of the Absolutist State* (London, 1974).

Anweiler, Oskar, *The Soviets: The Russian Workers, Peasants, and Soldiers Councils, 1905-1921*, trans. R. Hein (New York, 1974).

Astrov, N.I. and Gronsky, Paul P., *The War and the Russian Government* (New Haven, Conn., 1929).

Avrekh, A.Ia., 'Stolypinskii bonapartizm i voprosy voennoi politiki v III dume', *Voprosy istorii*, no. 11 (1956) pp. 17-33.

Bailes, Kendall E., *Technology and Society under Lenin and Stalin: Origins of the Soviet Technical Intelligentsia* (Princeton, N.J., 1978).

Balabanov, M., *Ot 1905 k 1917: Massovoe rabochee dvizhenie* (Moscow and Leningrad, 1927).

Barsukov, E.Z., 'Grazhdanskaia promyshlennost' v boevom snabzhenii armii', *Voina i revoliutsiia*, no. 10 (1928) pp. 11-22.

Barsukov, E.Z., *Podgotovka Rossii k voine v*

artilleriiskom otnoshenii (Moscow and Leningrad, 1926).

Bill, Valentine T., *The Forgotten Class: The Russian Bourgeoisie to 1900* (New York, 1959).

Blackwell, Wm L., 'The Old Believers and the Rise of Private Industrial Enterprise in Early Nineteenth Century Moscow', in *Russian Economic Development from Peter the Great to Stalin*, ed. Wm L. Blackwell (New York, 1976) pp. 138–58.

Borisov, S.P., *Bor'ba bol'shevikov protiv voenno-promyshlennykh komitetov* (Moscow, 1948).

Bovykin, V.I., 'Banki i voennaia promyshlennost' Rossii nakanune pervoi mirovoi voiny', *Istoricheskie zapiski*, vol. LXIV (1959) pp. 82–135.

Bovykin, V.I. and Shatsillo, K.F., 'Lichnye unii v tiazheloi promyshlennosti nakanune pervoi mirovoi voiny', *Vestnik Moskovskogo universiteta, seriia ix: istoriia*, no. 1 (1962) pp. 55–74.

Bovykin, V.I. and Tarnovskii, K.N., 'Kontsentratsiia proizvodstva i razvitie monopolii v metalloobrabatyvaiushchei promyshlennosti Rossii', *Voprosy istorii*, no. 2 (1957) pp. 19–31.

Burdzhalov, E.N., *Vtoraia russkaia revoliutsiia: Moskva, front, periferiia* (Moscow, 1971).

Chermenskii, E.D., *Burzhuaziia i tsarizm v revoliutsii 1905-1907 gg.* (Moscow and Leningrad, 1939).

Chernov, Victor, *The Great Russian Revolution*, trans. P. Mosely (New York, 1966).

Crisp, Olga, 'Labour and Industrialization in Russia', in *The Cambridge Economic History of Europe*, vol. VII, part 2 (London, 1978) pp. 308–415.

Crisp, Olga, *Studies in the Russian Economy before 1914* (London, 1976).

Dan, Theodore, *The Origins of Bolshevism*, trans. J. Carmichael (New York, 1970).

Danilov, You., *La Russie dans la Guerre mondiale, 1914-1916* (Paris, 1927).

Diakin, V.S., 'Chto takoe Progressivnyi blok?', *Voprosy istorii*, no. 1 (1970) pp. 200–4.

Diakin, V.S., *Russkaia burzhuaziia i tsarizm v gody pervoi mirovoi voiny, 1914-1917* (Leningrad, 1967).

Dvinov, B., *Pervaia mirovaia voina i Rossiiskaia*

Sotsial-demokratiia (New York, 1962).

Fallows, Thomas, 'Politics and the War Effort in Russia: the Union of Zemstvos and the Organization of the Food Supply, 1914-1916', *Slavic Review*, vol. XXXVII (1978) pp. 70-90.

Feldman, G.D., *Army, Industry, and Labor in Germany, 1914-1918* (Princeton, N.J., 1966).

Ferro, Marc, *La Revolution de 1917* (Paris, 1967).

Gaponenko, L.S., *Rabochii klass Rossii v 1917 godu* (Moscow, 1970).

Gefter, M.Ia., 'Toplivno-neftianoi golod v Rossii i ekonomicheskaia politika tret'ei iiunskoi monarkhii', *Istorichekie zapiski*, vol. LXXXIII (1969) pp. 76-122.

Gefter, M.Ia., 'Tsarizm i monopolisticheskii kapital v metallurgii Iuga Rossii', *Istoricheskie zapiski*, vol. XLIII (1953) pp. 70-130.

Gefter, M.Ia., 'Tsarizm i zakonodatel'noe "regulirovanie" deiatel'nosti sindikatov i trestov v Rossii nakanune pervoi mirovoi voiny', *Istoricheskie zapiski*, vol. LIV (1954) pp. 170-93.

Gindin, I.F., 'Problemy istorii Fevral'skoi revoliutsii i ee sotsial'no-ekonomicheskykh predposilok', *Istoriia SSSR*, no. 4 (1967) pp. 30-49.

Gindin, I.F., 'Russkaia burzhuaziia v period kapitalizma, ee razvitie i osobennosti', *Istoriia SSSR*, no. 2 (1963) pp. 57-89; no. 3 (1963) pp. 37-60.

Gindin, I.F., *Russkie kommercheskie banki* (Moscow, 1948).

Gindin, I.F. and Shepelev, L.E., 'Bankovie monopolii v Rossii nakanune velikoi Oktiabr'skoi sotsialisticheskoi revoliutsii', *Istoricheskie zapiski*, vol. LXVI (1960) pp. 21-95.

Girault, René, *Emprunts russes et Investissements français en Russie, 1887-1914* (Paris, 1973).

Girault, René, 'Finances internationales et relations internationales (à propos des usines Poutiloff)', *Revue d'histoire moderne et contemporaine*, vol. XIII (1966) pp. 217-36.

Golovine, N.N., *The Russian Army in the World War* (New Haven, Conn., 1931).

Gourko, B., *War and Revolution in Russia, 1914-1917* (New York, 1919).

Grave, B.B., *K istorii klassovoi bor'by v Rossii v gody imperialisticheskoi voiny* (Moscow and

Leningrad, 1926).

Grave, B.B., 'Militarizatsiia promyshlennosti i rossiiskii proletariat v gody pervoi mirovoi voiny', in *Iz istorii rabochego klassa i revoliutsionnogo dvizheniia, Sbornik* (Moscow, 1958) pp. 405-26.

Grunt, A.Ia., *Moskva 1917: Revoliutsiia i kontrrevoliutsiia* (Moscow, 1976).

Grunt, A.Ia., 'Progressivnyi blok', *Voprosy istorii*, no. 3-4 (1945) pp. 108-17.

Haimson, Leopold, 'The Problem of Social Stability in Urban Russia, 1905-1917', *Slavic Review*, vol. XXIII (1964) pp. 619-42; vol. XXIV (1965) pp. 1-22.

Hurwitz, S.J., *State Intervention in Great Britain: A Study of Economic Control and Social Response, 1914-1919* (New York, 1949).

Ipat'ev, V.N., *Rabota khimicheskoi promyshlennosti na oboronu vo vremia voiny* (Petrograd, 1920).

Ivanova, N.A., 'Prinuditelnye ob"edineniia v Rossii v gody pervoi mirovoi voiny', in *Ob osobennostiakh imperializma v Rossii* (Moscow, 1963) pp. 234-49.

Karasov, G.G., 'Bor'ba proletariata Moskvy protiv "rabochei gruppy" Moskovskogo oblastnogo voenno-promyshlennogo komiteta', *Vestnik Moskovskogo universiteta, seriia ix: istoriia*, no. 3 (1974) pp. 34-45.

Karatygin, P., *Obchchie osnovye mobilizatsii promyshlennosti* (Moscow, 1925). •

Kareev, K., 'Rabochaia pechat' i voenno-promyshlennye komitety', *Krasnaia letopis'*, no. 21 (1926) pp. 13-52.

Kaser, Michael C., 'Russian Entrepreneurship', in *The Cambridge Economic History of Europe*, vol. VII, part 2 (London, 1978) pp. 416-93.

Katkov, George, *Russia 1917: The February Revolution* (London, 1969).

Kats, A., 'Bor'ba s bezrabotitsei v gody voiny', *Materialy po istorii professional'nogo dvizheniia v Rossii*, no. 3 (1925) pp. 60-87.

Keep, John L.H., *The Russian Revolution: A Study in Mass Mobilization* (New York, 1976).

Kharitonov, V.L., 'Bor'ba bol'shevikov Ukrainy za boikot voenno-promyshlennykh komitetov', *Nauchnye doklady vyshei shkoly istoricheskoi nauki*, no. 44 (1958) pp. 55-67.

Kir'ianov, Iu.I., *Rabochie Iuga Rossii, 1914-*

fevral' 1917 g. (Moscow, 1971).

Kovalenko, D.A., *Oboronnaia promyshlennost' sovetskoi Rossii v 1918-20 gg.* (Moscow, 1970).

Krupina, T.D., 'K voprosu o vzaimootnosheniiakh tsarskogo pravitel'stva s monopoliiami', *Istoricheskie zapiski*, vol. LVII (1956) pp. 144-76.

Krupina, T.D., 'Politicheskii krizis 1915 g. i sozdanie Osobogo soveshchaniia po oborone', *Istoricheskie zapiski*, vol. LXXXIII (1969) pp. 58-75.

Kruze, E.E., 'Rabochie Peterburga v gody novogo revoliutsionnogo pod"ema', in *Istoriia rabochikh Leningrada*, 2 vols (Leningrad, 1972) vol. I, pp. 449-60.

Laverychev, V.Ia., *Monopolisticheskii kapital v tekstil'noi promyshlennosti Rossii, 1900-1917* (Moscow, 1963).

Laverychev, V.Ia., *Po tu storonu barrikad* (Moscow, 1967).

Laverychev, V.Ia., 'Protsess monopolizatsii khlopchatobumazhnoi promyshlennosti Rossii (1900-1914 gody)', *Voprosy istorii*, no. 2 (1960) pp. 137-61.

Laverychev, V.Ia., *Tsarizm i rabochii vopros v Rossii, 1861-1917 gg.* (Moscow, 1972).

Laverychev, V.Ia., 'Vserossiiskii soiuz torgovli i promyshlennosti', *Istoricheskie zapiski*, vol. LXX (1961) pp. 35-60.

Leiberov, I.P., 'Deiatel'nost' Petrogradskoi organizatsii bol'shevikov i ee vliianie na rabochee dvizhenie v Rossii v gody pervoi mirovoi voiny', in *Pervaia mirovaia voina*, ed. A.L. Sidorov (Moscow, 1968) pp. 283-302.

Leiberov, I.P., 'O vozniknovenii revoliutsionnoi situatsii v Rossii v gody pervoi mirovoi voiny (iiul'-sent. 1915 g.)', *Istoriia SSSR*, no. 6 (1964) pp. 33-59.

Leiberov, I.P., 'Petrogradskii proletariat v gody pervoi mirovoi voiny', in *Istoriia rabochikh Leningrada*, 2 vols (Leningrad, 1972) vol. I, pp. 461-511.

Leiberov, I.P., 'Vtoroi den' Fevral'skoi revoliutsii (sobytiia 24 fevralia 1917 g. v Petrograde)', in *Sverzhenie samoderzhavia: Sbornik statei* (Moscow, 1970) pp. 100-19.

Liashchenko, P.I., 'Iz istorii monopolii v Rossii', *Istoricheskie zapiski*, vol. XX (1946) pp. 150-88.

Livshin, Ia.I., *Monopolii v ekonomike Rossii* (Moscow, 1961).

Livshin, Ia.I., 'Predstavitel'nye organizatsii krupnoi burzhuazii v Rossii v kontse XIX-nachale XX vv.', *Istoriia SSSR*, no. 2 (1959) pp. 95-117.

Maevskii, I.V., *Ekonomika russkoi promyshlennosti v usloviiakh pervoi mirovoi voiny* (Moscow, 1957).

McKay, John P., *Pioneers for Profit: Foreign Entrepreneurship and Russia Industrialization, 1885-1913* (Chicago, 1970).

Mel'gunov, S.P., *Na putiakh k dvortsovomu perevorotu* (Paris, 1931).

Mendel, Arthur, 'On Interpreting the Fate of Imperial Russia', in *Russia under the Last Tsar*, ed. Theofanis G. Stavrou (Minneapolis, 1969) pp. 13-41.

Mints, I.I., 'Revoliutsionnaia bor'ba proletariata Rossii v 1914-1916 godakh', *Voprosy istorii*, no. 11 (1959) pp. 57-70.

Owen, Thomas C., *Capitalism and Politics in Russia: A Social History of the Moscow Merchants, 1855-1905* (Cambridge, 1981).

Pares, B., *The Fall of the Russian Monarchy* (New York, 1961).

Pearson, Raymond, *The Russian Moderates and the Crisis of Tsarism, 1914-1917* (New York, 1977).

Plugatyrev, P.G., 'Bor'ba bol'shevistskoi organizatsii Khar'kova protiv voenno-promyshlennykh komitetov v period pervoi mirovoi voiny', *Sbornik nauchnye rabot kafedr istorii KPSS vuzov g. Khar'kova*, no. 3 (1960) pp. 35-42.

Pogrebinskii, A.P., *Gosudarstvenno-monopolisticheskii kapitalizm v Rossii* (Moscow, 1959).

Pogrebinskii, A.P., 'K istorii soiuzov zemstv i gorodov v gody imperialisticheskoi voiny', *Istoricheskie zapiski*, vol. XII (1941) pp. 39-60.

Pogrebinskii, A.P., 'Mobilizatsiia promyshlennosti tsarskoi Rossii v pervuiu mirovuiu voinu', *Voprosy istorii*, no. 8 (1948) pp. 58-70.

Pogrebinskii, A.P., 'Sindikat "Prodamet" v gody pervoi mirovoi voiny, 1914-1917', *Voprosy istorii*, no. 10 (1958) pp. 22-34.

Pogrebinskii, A.P., 'Voenno-promyshlennye komitety', *Istoricheskie zapiski*, vol. XI (1941) pp. 160-200.

Portal, Roger, 'Muscovite Industrialists: the

Cotton Sector, (1861-1914)', in *Russian Economic Development from Peter the Great to Stalin*, ed. Wm L. Blackwell (New York, 1976) pp. 161-96.

Rashin, A.G., *Formirovanie rabochego klassa Rossii* (Moscow, 1958).

Reikhardt, V.V., 'K probleme monopolisticheskogo kapitalizma v Rossii – evoliutsiia uchastiia inostrannogo kapitala v russkom narodnom khoziaistve za gody imperialisticheskoi voiny, 1914-1917', *Problemy marksizma*, no. 5-6 (1931) pp. 193-222.

Riha, Thomas, 'Milyukov and the Progressive Bloc in 1915: a Study in Last Chance Politics', *Journal of Modern History*, vol. XXXII (1960) pp. 16-24.

Romanov, F., 'Voenno-promyshlennye komitety i boikot ikh bol'shevikami', *Profsoiuzy SSSR*, no. 6 (1940) pp. 42-56.

Roosa, Ruth A., 'Russian Industrialists and "State Socialism", 1906-17', *Soviet Studies*, vol. XXIII (1972) pp. 395-417.

Roosa, Ruth A., 'Russian Industrialists Look to the Future: Thoughts on Economic Development, 1906-1917', in *Essays in Russian and Soviet History*, ed. John S. Curtiss (Leiden, 1963) pp. 198-218.

Roosa, Ruth A., '"United" Russian Industry', *Soviet Studies*, vol. XXIV (1973) pp. 421-5.

Rozental', I.S., 'Russkii liberalizm nakanune pervoi mirovoi voiny i taktika bol'shevikov', *Istoriia SSSR*, no. 6 (1971) pp. 52-70.

Seiranian, B.S., *Bor'ba bol'shevikov protiv voenno-promyshlennykh komitetov* (Erevan, 1961).

Semennikov, V.P., *Monarkhiia pered krusheniem, 1914-1917* (Moscow and Leningrad, 1927).

Shatsillo, K.F., 'O disproportsii v razvitii vooruzhennykh sil Rossii nakanune pervoi mirovoi voiny (1906-1914 gg.)', *Istoricheskie zapiski*, vol. LXXXIII (1969) pp. 123-36.

Shmukker, M.M., *Ocherki finansov i ekonomiki zheleznodorozhnogo transporta Rossii za 1913-1922 gody* (Moscow, 1923).

Sidorov, A.L., 'Bor'ba s krizisom vooruzheniia russkoi armii v 1915-1916 gg.', *Istoricheskie zhurnal*, no. 10-11 (1944) pp. 35-57.

Sidorov, A.L., *Ekonomicheskoe polozhenie Rossii v gody pervoi mirovoi voiny* (Moscow, 1973).

Sidorov, A.L., 'Evakuatsiia russkoi promy

shlennosti vo vremia pervoi mirovoi voiny', *Voprosy istorii*, no. 6 (1947) pp. 3-25.

Sidorov, A.L., *Finansovoe polozhenie Rossii v gody pervoi mirovoi voiny* (Moscow, 1960).

Sidorov, A.L., 'K voprosu o stroitel'stve kazennykh voennykh zavodov v Rossii v gody pervoi mirovoi voiny', *Istoricheskie zapiski*, vol. LIV (1954) pp. 156-69.

Sidorov, K.F., 'Rabochee dvizhenie v Rossii v gody imperialisticheskoi voiny, 1914-1917 gg.', in *Ocherki po istorii Oktiabr'skoi revoliutsii*, ed. M.N. Pokrovskii, 2 vols (Moscow and Leningrad, 1927) vol. 1, pp. 342-421.

Solov'ev, Iu.B., 'Protivorechiia v praviashchem lagere Rossii po voprosu ob inostrannykh kapitalakh v gody pervogo promyshlennogo pod"ema', in *Iz istorii imperializma v Rossii* (Moscow, 1959) pp. 371-88.

Somov, S.A., 'O "maiskom" Osobom soveshchanii', *Istoriia SSSR*, no. 3 (1973) pp. 112-23.

Startsev, V.I., *Russkaia burzhuaziia i samoderzhavie v 1905-1917 gg. (Bor'ba vokrug 'otvetstvennogo ministerstva' i 'Pravitel'stva doveriia')* (Leningrad, 1977).

Stone, Norman, *The Eastern Front, 1914-1917* (London, 1975).

Strumilin, S.G., *Ocherki ekonomicheskoi istorii Rossii* (Moscow, 1960).

Tarnovskii, K.N., *Formirovanie gosudarstvenno-monopolisticheskogo kapitalizma v Rossii v gody pervoi mirovoi voiny* (Moscow, 1958).

Tarnovskii, K.N., 'Komitet po delam metallurgicheskoi promyshlennosti i monopolisticheskie organizatsii', *Istoricheskie zapiski*, vol. LVII (1956) pp. 80-143.

Tarnovskii, K.N., *Sovetskaia istoriografiia Rossiiskogo imperializma* (Moscow, 1964).

Tiutiukin, S.V., *Voina, mir, revoliutsiia: Ideinaia bor'ba v rabochem dvizhenii Rossii, 1914-1917 gg.* (Moscow, 1972).

Tolf, Robert W., *The Russian Rockefellers: The Saga of the Nobel Family and the Russian Oil Industry* (Stanford, Calif., 1976).

Trotsky, L., *The History of the Russian Revolution*, trans. M. Eastman, 3 vols (London, 1967).

Tsukernik, A.L., *Sindikat 'Prodamet'* (Moscow, 1959).

Tsvibak, M.M., *Iz istorii kapitalizma v Rossii, khlopchatobumazhnaia promyshlennost'* (Leningrad, 1925).

Volobuev, P.V., 'Iz istorii sindikata "Produgol'"', *Istoricheskie zapiski*, vol. LVII (1956) pp. 107-41.

Volobuev, P.V., 'Politika proizvodstva ugol'nykh i neftianykh monopolii v Rossii nakanune pervoi mirovoi voiny', *Vestnik Moskovskogo universiteta, seriia ix: istoriia*, no. 1 (1956) pp. 71-115.

Voronkova, S.V., 'Stroitel'stvo avtomobil'nykh zavodov v Rossii v gody pervoi mirovoi voiny (1914-1917 gg.)', *Istoricheskie zapiski*, vol. LXXV (1965) pp. 147-69.

Vovchik, A.F., *Politika tsarizma po rabochemu voprosu v predrevoliutsionnyi period, 1895-1904* (L'vov, 1964).

White, James D., 'Moscow, Petersburg and the Russian Industrialists', *Soviet Studies*, vol. XXIV (1973) pp. 414-20.

Wildman, Allan K., *The End of the Russian Imperial Army: The Old Army and the Soldiers' Revolt (March-April, 1917)* (Princeton, N.J., 1980).

Yatsunsky, V.K., 'The Industrial Revolution in Russia', in *Russian Economic Development from Peter the Great to Stalin*, ed. Wm L. Blackwell (New York, 1974) pp. 111-35.

Zagorsky, S.O., *State Control of Industry in Russia during the War* (New Haven, Conn., 1928).

Zaionchkovskii, P.A., *Samoderzhavie i russkaia armiia na rubezhe xix-xx stoletii, 1881-1903* (Moscow, 1973).

Index

Printed in Great Britain
by Amazon

48109466R00193